Harvard East Asian Series 80

The East Asian Research Center at Harvard University
administers research projects
designed to further scholarly understanding
of China, Japan, Korea, Vietnam, Inner Asia,
and adjacent areas.

Toward Industrial Democracy

Management and Workers in Modern Japan

KUNIO ODAKA

Harvard University Press
Cambridge, Massachusetts

To John C. Pelzel

In remembrance of years of friendship

Preface

Japan's remarkable economic growth and industrial progress since the end of World War II are common knowledge. Westerners have not known at all well, however, the lives and minds of the people—workers and managers—who have created this unique development. I do not intend to disparage the writings of Western scholars on management practices and working conditions in Japanese industry, but I believe that most give an inadequate understanding of the phenomena, above all perhaps because of what I consider to be misinterpretations.

In this little book I have tried to show how managerial ideologies and practices have changed, how the workers have reacted to recent technological advances, how they have thought of their companies and labor unions, what their motivations for work have been, and the meaning to them of their leisure time. Beyond this, I have given special attention to a very significant trend that has come into evidence in recent years, a trend toward the democratization of Japanese industrial management. In addition to a description of this trend, I have ventured to offer what I consider to be a feasible program for promoting industrial democracy in my country.

Chapter One, "Traditionalism and Democracy in Japanese Industry," introduces the problems and perspectives as I see them. I have in fact here, through a critique of certain theories about the Japanese pattern of management advanced by Western scholars, tried to suggest the existence of a recent evolution from traditionalism to democracy in Japanese management.

The four chapters that follow present data and interpretations on the conditions and attitudes of Japanese workers taken from a number of empirical studies which my collaborators and I conducted. Spanning almost all the years from the very beginning of recovery after the devastation of the war through to the attainment by Japan of the position of the world's third economic power, these studies have focused upon such matters as the workers' experience with their jobs, the physical environment and human relations of their workshops, their working conditions and leisure activities, changes in technology and productivity in their companies, and their firms' managerial policies, labor–management relations, and their labor unions. Of particular interest are several series of studies—one, conducted over the years from 1952 to 1967 in nine large Japanese companies, on the "allegiance" of workers to both their company and their union; another, undertaken at two firms in 1966 and 1967, of employee attitudes toward work and leisure; and two surveys, carried out under the sponsorship of the Japanese Ministry of Labor with a nationwide sample of industrial plants in 1967 and 1968, on workers' experience of "monotonous" jobs. I believe that these studies clarify worker consciousness and attitudes and their correlates better than any other material and show better the evolution of workers' points of view across the entire span of Japan's recent economic expansion.

Based upon this material, in Chapter Two I have examined worker reactions to recent technological changes, in Chapter Three the opportunity that technological innovation has given younger employees to display their real abilities, in Chapter Four the workers' identification with union and company, and in Chapter Five their attitudes toward work and leisure.

The final chapter is of a different nature. It is in fact programmatic and presents measures which I believe should be taken to democratize—and I would therefore say modernize—Japanese industrial relations. In my view, the crux of this matter is the introduction and development of a system of workers' participation in management and of their self-government in the workshop. Such a departure would relieve work monotony and the sense of estrangement from the work environment that have been growing in recent years; it would at the same time improve the effectiveness and productivity of the industrial organization itself.

The program that I propose in Chapter Six has of course grown out of my empirical work, but it was first formulated in an article I intended mainly for company executives, published in a leading general magazine, *Chuo Koron,* in May 1965. Since then, I have met many times in lectures and seminars with company executives, personnel specialists, union leaders, and social scientists concerned with these matters, and my argument here has gained much from their comments and suggestions, which I have tried to examine and incorporate into the present English version. In view of the fact that there has been a revival of American interest in democratic-participative systems of management in their own affairs, I hope that this chapter will be not only a program for Japanese management but a source of stimulation for American management as well.

Chapters included in this volume were originally prepared on different occasions. Chapters One and Five were first written in English as my contributions to the Fifth and Sixth World Congresses of Sociology, held in Washington, D.C., and in Evian, France, in 1962 and 1966. The rest were originally written in Japanese. These articles, including Japanese versions of Chapters One and Five, were incorporated into my two recent books, *Nihon no keiei* (Japanese industrial management) and *Shokugyo no rinri* (Work ethics), published by the Chuo Koron Publishing Company in 1965 and 1970, respectively. In preparing this English edition, I have tried to coordinate the chapters so as to form an integrated whole, and most chapters have been both rewritten and enlarged by supplementary data and fresh interpretations.

The work of preparing the English edition was carried out during the period from July 1970 to February 1972, when I was given a chance to work as a research fellow at the East Asian Research Center of Harvard University, on a grant made jointly by that Center and the Harvard–Yenching Institute. Most of the time from November 1970 through December 1971 I spent in collaboration with John C. Pelzel, Director of the Harvard–Yenching Institute and Professor of Anthropology at Harvard, he taking the role of editor for my manuscript, while I took that of adviser to him in the preparation of chapters on Japanese culture for a work he had in progress. I am deeply indebted to Pelzel, who during this period willingly spared his precious time to improve my English. Professor Alex Inkeles, the distinguished sociologist now at Stanford, who had read my earlier papers and a draft of the present book, gave me continual support and encouragement throughout the preparation of the volume. Professor Albert Craig, Associate Director of the East Asian Research Center, was kind enough to read the manuscript and give me constructive criticism and helpful advice. Without the assistance of these friends, this book would never have been completed in its present form. I must also express my debt to the East Asian Research Center for its decision to include this book under its aegis.

Finally, I would like to express my gratitude to Kyoko Odaka, my wife, for her unflagging aid and encouragement during the period I was preparing the English manuscript. I shall never forget the days and nights I spent with her working in a small, white New England house, surrounded by spacious lawns and beyond them by a quiet and lovely woodland, in Westwood, twenty miles away from Harvard.

Kunio Odaka

Contents

xiii

Tables

Toward Industrial Democracy

Traditionalism and Democracy

in Japanese Industry

Preindustrial Practices in Industrial Japan

In spite of the fact that Japan is one of the world's most highly industrialized nations, a variety of preindustrial practices seems still to exist in its present-day industrial relations and managerial policies. These practices include, among others, "lifetime" employment relationships, a wage and promotion system, often referred to as the *nenko* system, which is based primarily on age and length of service, almost a class distinction between superior and subordinate, great stress placed on personal loyalty toward the employer rather than on work efficiency, emphasis on intragroup harmony rather than on individual competition, and paternalistic care of employees, even in their private lives. The social forces working to maintain these practices may be called "traditionalism."

Two contrasting views have been advanced regarding traditionalism in Japanese industry. One of them argues that these practices are nothing but carry-overs from the feudalistic relations between lord and retainer and the relations of the patriarchal family. Industrialization in Japan was a process initiated and promoted primarily by the government, rather than a movement which arose from among the people, as in Western countries. Again, the early leaders of industrialization came from the patriotic *samurai* classes, who were accustomed to the traditional value orientations of feudal Japan. As a consequence, managerial policies and industrial relations in Japan were from the beginning, and are even today, tinged with various premodern traits.

According to this theory, industrialization and economic development progress in accord with a uniform pattern everywhere throughout the world, much like the alleged evolutionary development from the feudal to the capitalist, and from the capitalist to the socialist, order of society. The fact that the traditional practices are still in force, therefore, points to a disreputable backwardness which is an obstacle to the democratization and modernization of Japanese industry.

This theory, which may be called the theory of "uniform development," asserts that the harmful effects of traditionalism are readily apparent. The system of lifetime employment, for example, leads to overstaffing and makes it difficult to discharge incompetent employees, resulting in a general reduction of efficiency. Since permanent or regular employees may not be discharged, many companies use a considerable number of "temporary" workers, who can be laid off in time of depression, and this practice results in undemocratic discrimination between them and the regular workers. The *nenko* system, on the other hand, discourages able young workers from exerting their talents and abilities. Traditionalism also accounts for the prevalence of "enterprise unions," unions whose members are confined to a single enterprise, and which are often strongly dependent on management.

What is worse, proponents of this theory continue, corresponding preindustrial traits often characterize the workers themselves. The majority of Japanese workers, for example, until recently were recruited from rural districts and were basically accustomed to the preindustrial ways and values of rural patriarchal families. As a result, they are even today apt uncritically to accept the patriarchal authority of management, and while loyal to their company, they tend to rely too much upon it.

Thus, supporters of this theory argue that if Japanese industry is to be democratized and modernized, the traditional practices, and particularly the system of lifetime employment, must be eliminated, and labor unions must come to organize on an industrial rather than an enterprise basis. The theory of uniform development was a popular view among Japanese social scientists—especially those with left-wing inclinations—until quite recently.[1]

The other opinion, which may be called the theory of "pluralistic industrialism," has been advocated since the 1950s, mostly by American labor economists and sociologists, like Solomon B. Levine, James C. Abegglen, Frederick H. Harbison, Charles A. Myers, and Clark Kerr.[2] According to this theory, the existence of traditional practices does not necessarily mean backwardness, even though they may be a remnant of the preindustrial stage, nor are they obstacles to

1. See, for example, Nihon Jinbun Kagakkai (Japan Humanistic Science Society), *Hoken isei* (Feudalistic remnants; Tokyo: Yuhikaku, 1951); Kazuo Okochi, ed., *Nihon no rodo kumiai* (Labor unions in Japan; Tokyo: Toyo Keizai Shinpo-sha, 1954); Shakai Seisaku Gakkai (Society for the Study of Social Policy), *Chin rodo ni okeru hokensei* (Feudalistic practices in wage labor relations; Tokyo: Yuhikaku, 1955).

2. Solomon B. Levine, *Industrial Relations in Postwar Japan* (Urbana: University of Illinois Press, 1958); James C. Abegglen, *The Japanese Factory: Aspects of Its Social Organization* (Glencoe, Ill.: Free Press, 1958); Frederick H. Harbison and Charles A. Myers, *Management in the Industrial World: An International Analysis* (New York: McGraw-Hill, 1959); Clark Kerr, John T. Dunlop, Frederick H. Harbison, and Charles A. Myers, *Industrialism and Industrial Man: The Problems of Labor and Management in Economic Growth* (Cambridge, Mass.: Harvard University Press, 1960).

the modernization of Japanese industry. Rather, they in fact aided the unusually rapid and successful development of Japanese industry during the period from the late nineteenth to the early twentieth century. They have also contributed to the reconstruction and prosperity of Japanese industry since the end of World War II.

Generally speaking, this theory asserts, the processes of industrialization and economic growth will vary from country to country in accordance with the differences in the political situation, social structure, way of life, value system, and so forth, of the country concerned.[3] The very factors that promote industrial progress in one country, therefore, may obstruct it in another, and vice versa.

It has often been said, for instance, that the Protestant ethic, the competitive and individualistic nature of social interaction, and the rationalistic character of industrial organization contributed constructively to the industrial productivity and economic growth of Western countries. In non-Western countries, including Japan, however, the same factors may not be useful. By contrast, from the viewpoint of Western industrialism, the nationalistic motivation of entrepreneurs, the immobility of the work force, personal loyalty to the employer, and paternalistic care for the welfare of employees may seem to be inefficient and feudalistic. Nevertheless, in terms of the specific Japanese social and cultural background, these factors did foster industrial progress and did raise the national standard of living.[4]

This peculiarity of Japan's situation, according to the pluralistic theory, is closely related to the fact that in Japan

3. See Levine, *Industrial Relations*, pp. viii–ix; Clark Kerr et al., *Industrialism and Industrial Man*, pp. 15–32.

4. Everett E. Hagen, in the foreword to Abegglen's book, laconically remarks as follows: "Indeed, if judged from the viewpoint of the American business executive, Japanese personnel principles and relationships are inefficient. Yet they are efficient—highly efficient. The proof of the pudding is in the eating." Abegglen, *Japanese Factory*, p. vii.

4

the entrepreneurs who guided the initial stage of industrialization were "dynastic elites," who brought to their new role feudalistic life styles and ways of thinking. In the West, as typified by England, the early industrial leaders were a new middle class who had new value systems and ways of life. Under their leadership, the traditional hierarchical elements of society were rapidly eliminated, and new patterns of industrial organization and democratic culture became widespread. In the case of Japan, on the other hand, the leadership by dynastic elites necessarily led to the preservation of such practices as a hierarchical employer–employee relationship, on the model of the feudal master and servant, and personnel management emphasizing submission rather than efficiency, years of service as opposed to achievement, and harmony instead of competition.

The traditional social atmosphere thus preserved within the enterprise, in its turn, formed an ideal environment for the workers of those times, who had similar life styles and were able to fit themselves to the new establishment with comparative ease. If it had not been for this climate, workers without experience in modern technology, the use of mechanized equipment, and the restrictive regulations of the work place would have suffered such a shock that their will to produce would have been lessened, or, as happened in many industrializing countries of the West, they might have resisted the change violently. Fortunately or unfortunately, however, the continuation of a premodern social environment eased the workers' transition, so that entrepreneurs were able steadily to increase work productivity through a "stick and carrot," or autocratic but paternalistic, form of management. Although wages were low and working conditions severe, the workers, who had been uprooted from their native villages, were given what might be called a "second home."

In short, in Western countries industrial development was

5

carried out by breaking up feudalistic and traditional practices, but in Japan it has been achieved by preserving, and to a certain extent by utilizing, them. Industrialization in Japan is thus unique in that it has been carried out without destroying social and cultural continuity. According to Abegglen, the Japanese experience should be classified as a third pattern of industrialization, quite different from that of either Western or Communist countries.[5]

Pitfalls in the Theory of Pluralistic Industrialism

While the theory of uniform development adheres to the formulistic view that industry grows according to the same pattern everywhere in the world, regardless of social and cultural differences, the pluralistic theory emphasizes the peculiar nature of industrialization in each country. Whereas the former criticizes traditionalism in Japanese industry by pointing out its obstructive role, the latter develops a new view stressing its positive contributions. Which of these two contrasting theories, then, is more in accord with reality? My opinion is that both contain exaggerations and misinterpretations and, therefore, require reconsideration.

There are of course several merits in the theory of pluralistic industrialism. As a basis for explanation it is more fruitful, and as an analytical framework it is more realistic, than the theory of uniform and unilinear development which has predominated in the past.

According to the theory of uniform development, industrialization and economic growth in all countries can be explained in terms of a single measure, consisting of a set of universal factors, such as the degree of capital accumulation, the progress of technology, the size of the labor force, and

5. *Ibid.*, pp. 1–2.

the productivity rate. In actuality, industrialization has at least two aspects: a universal and a particular. The former consists of those factors which are common to all cases, and which can be, at least to a certain extent, quantified, such as those just enumerated. In contrast, the particular aspect comprises those factors peculiar to a given country and hardly capable of quantification, such as its class structure, its family system, its religion, its social customs and norms, and the value orientations of its people. The degree of industrialization as measured by the universal factors will necessarily be affected by those which are particular, and which are by no means of the same nature everywhere. It was for this very reason that Clark Kerr and other American labor economists advocated the theory of pluralistic industrialism.

However, there are also pitfalls in this new approach. For one thing, it tends to overstress the particular aspects of industrialization, and in the Japanese case is even based upon some misinterpretation of the facts. For example, Abegglen, in his *The Japanese Factory,* treats as the "critical difference" between Japanese and Western industrial organizations the system of "lifetime" employment,[6] which he and some other scholars consider to be a feudalistic remnant of the Tokugawa period.[7] Actually, prior to the 1920s, labor mobility was fairly high even in large enterprises, and especially in the small and often technologically more primitive enterprises which have until very recently comprised a much larger portion of the total industrial establishment in Japan than in the West.[8] Similarly, at present the rate of labor

6. *Ibid.,* pp. 11–25.
7. *Ibid.,* pp. 130–135; Levine, *Industrial Relations,* pp. 32–35.
8. Statistics compiled from recent census data show that the proportion of minor enterprises employing fewer than 100 workers, where labor turnover is much more frequent, was 61 per cent of all enterprises in Japan in the late 1950s, while the corresponding proportion for the United States, Great Britain, and West Germany was only about 25 per cent each. See Nihon Seisansei Honbu (Japan Productivity Center), *Nihon no chusho kigyo* (Minor enterprises in Japan; Tokyo: Nihon Seisansei Honbu, 1958).

mobility in Japanese industry is again rising.[9] It was only between the end of World War I and the very recent period of rapid business expansion that some sectors of Japanese industry—usually the larger and to workers more attractive—developed to an extreme degree the pattern of relatively long terms of service that Abegglen describes. His conclusions, while to a certain extent true, are exaggerated, neglecting the fluctuations in the history of Japanese industrial development just noted, and reflecting the practices of only a part of Japanese business firms.

For another thing, the pluralistic theory leads to the simplistic conclusion that "East is East and West is West," and this in its turn is all too often used in self-justification by conservative elements in Japanese business circles. Both some Japanese business leaders and some of those scholars who serve as their consultants are antipathetic to the formulistic theory of uniform development, which is often identified as Marxist, and favor the pluralistic theory, because of its opposition to the former, as well as because of its implied praise of traditional Japanese culture. In effect they say: "The essential point is making profits for your enterprise by maintaining a high rate of productivity. In relation to this objective, it does not matter whether traditional practices remain intact or industrial relations are democratized." Not infrequently, in other words, those with conservative inclinations

---

9. According to the Labor Turnover Survey conducted by the Japanese Ministry of Labor, for example, the rate of separation by regular workers in manufacturing industries during the year 1959 was 27.8 per cent in the case of small-scale enterprises employing 30 to 99 workers, as against 16.1 per cent and 6.7 per cent in larger firms with 100 to 499 workers and over 500, respectively. The separation rate for all manufacturing establishments amounted to 23.9 per cent, which was comparable to the corresponding ratio of 27.2 per cent in Great Britain in the same year. See Rodo-sho (Ministry of Labor), *Rodo ido chosa kekka hokoku* (Report of the labor turnover survey; Tokyo: Rodo-sho, 1960). The separation rates have gradually increased since then, and in 1964 that for all manufacturing industries came to 30.7 per cent, while that for small-scale enterprises with 30 to 99 employees reached 37.2 per cent. See Ministry of Labor, *Rodo hakusho* (Labor white paper; Tokyo: Rodo-sho, 1969).

use the theory of pluralistic industrialism as a justification for neglecting to democratize industrial relations or to introduce necessary administrative innovations into their companies.

Abegglen's book seems to have had a similar effect, by stressing the merits, rather than the demerits, of traditional practices. It effectively gave those who had tended to emphasize, from the viewpoint of uniform and unilinear development, the "backwardness" and "feudalistic remnants" in Japanese industrial management a chance to reconsider. At the same time, however, it gave the conservative elements among Japanese businessmen an excuse for maintaining self-righteousness and reactionary views. It is not farfetched to assume that this was one of the reasons that the book was enthusiastically welcomed in Japan.

Postwar Trends to Democracy

Perhaps the most serious drawback in the theory of pluralistic industrialism, however, is that it tends to overlook, or at least to underestimate, the rapid transformations that have been taking place in Japan since the end of World War II, and particularly since the latter half of the 1950s. These transformations are roughly in a direction that can be called "democratic" and on value as well as practical grounds must not be denied.

The factors that have brought about these trends include: (1) postwar labor legislation aimed at protecting the interests of wage earners, as represented by the Trade Union and the Labor Standards Laws; (2) a rapid spread of trade unionism since the end of the war; (3) the introduction of various technological innovations, such as automation, into Japanese industry; (4) changes in the attitudes of industrial workers, and especially of those of the younger generation, toward work and authority in industry—attitudes which have been influenced by the new educational system and by the at least partially democratic social atmosphere of postwar Japan; and (5) the effects of recent rapid economic growth, which has brought about an expansion of business, a shortage of skilled workers, increased labor mobility, and a general improvement of the national standard of living.

One of the major consequences has been a marked decline in the power of Japanese management, and it is in fact no longer accurate to characterize it as "authoritarian." Postwar labor legislation and the labor union movement have greatly restricted its power for coercive or arbitrary treatment of employees, and circumscribed its authority. Moreover, whereas most prewar workers came to the cities from rural districts, as the followers or protégés of successful older men in business, the postwar generation—mostly city-bred and heavily influenced by the democratic trends of postwar society and, as a result of the labor shortage, having opportunities for alternative employment—does not readily accept the patriarchal authority of an employer.

Another point which should not be underestimated is that such efforts as expanding welfare facilities or spending more money on personal fringe benefits, which management used to consider to be benevolent to employees, are no longer thought of as such by them. According to Levine, postwar employers in Japan have rejected "despotic tendencies inherent in the traditional approaches" but are resurrecting "managerial paternalism of the patriarchal type."[10] While it is true that there still remain some exceptionally conservative employers who stick to a traditionalistic approach, which demands allegiance to the company and high work morale in exchange for benevolence and paternalistic care, the chances that this strategy will succeed are rapidly decreas-

10. Levine, *Industrial Relations*, pp. 53–54.

ing. Today workers tend to think, quite wryly, that welfare facilities, and so on, are merely a cheap supplement for low wages, or provide things due to them anyway. It is clear, from the fact that recent labor strife has often taken the form of what are called "welfare struggles," that workers now consider the expansion of welfare facilities to be something they themselves have the right to fight for. A democratic spirit, fostered through the labor union movement, long ago enabled them to shake off much of their earlier feeling of indebtedness and loyalty to the company.

Along with the decline of management's authoritarian philosophy, there has been a gradual weakening of the traditionally sharp status distinctions between superior and subordinate. Technological innovation has robbed older forms of craftsmanship of the prestige they could preserve in a more slowly changing economy, and more generally the younger generation's respect for the authority of seniors seems to have declined.

Likewise, there is a growing tendency for Japanese management to attach importance to employees' actively exerting their individual talents, rather than maintaining the traditional attitude of passive obedience. Some progressive business executives have even come to realize the significance of encouraging employees' participation in managerial decisions.

The traditional system for determining wage increases and promotion primarily by seniority is also undergoing a change. Quite often management observes that younger and more highly skilled workers are not willing to work hard under the traditional seniority system, where less skilled but older employees are likely to be better treated. To meet this situation some firms have started to pay higher initial salaries than before on a general basis, but this measure can place a heavy burden on personnel costs, so long as the same firms maintain the traditional system, which continues the practice of periodical and uniform promotion. Thus many large enterprises are trying to revise the system of wages and promotions to one based primarily on job specification or efficiency rating.

The practice of lifetime employment has not been left untouched by these transformations. To be sure, the tendency for management to hold on to workers will continue, so long as the shortage of labor resulting from the current business expansion continues. Nevertheless, the remarkable economic growth in recent years has also forced management to favor an aggressive policy for recruiting necessary personnel even by luring them away from other firms. This new attitude will contribute toward the development of a free labor market in which workers can easily move from one enterprise to another without disadvantage. There is a corresponding tendency for workers readily to change their work place whenever they find jobs in which they can better themselves.

Although more examples of change may be cited, those given above should be sufficient to show that traditional practices, so greatly emphasized by proponents of the theory of pluralistic industrialism, are becoming less and less characteristic in recent years.

Toward Participative Management

In sum, Japanese industry is now showing signs of making progress toward an industrial democracy comparable to that found among non-Communist nations of the West. This evolution, however, is still in its earliest stages. While there is, for example, a definite tendency for the traditional *nenko* system to be replaced by a new method basing wages and promotions primarily on job specification and individual achievement, it may take a couple of decades for this shift to be realized for Japanese industry as a whole. Although there are business executives who have recently come to realize the significance of employee participation in managerial decisions, this important practice is still confined to a limited circle of progressive firms.

If the managerial policies of Japanese industry are to be modernized, and if they are to shake off the autocratic-paternalistic pattern, with any speed, therefore, purposeful efforts must be made to this end. Likewise, if management, in this era of technological innovation and of generally democratic trends, really wants to keep up the morale and motivation of its employees and to maintain high standards of work productivity, it must somehow reform its practices in a new and more democratic direction.

Workers will of course have to be fitted into such reformed practices, but it is the managers who must devise them or allow for their development and create an environment in which they will be viable. Management's innovative role here, as in other aspects of business, is crucial.

In Japan it is in fact often said that business executives have a public responsibility, that they manage a "public establishment" rather than private property, and that they should be conscious that they are the "elite of the country's economic growth." It is surely a good thing for them to realize that this obligation is theirs, and the role is clearly a better one than the pursuit of their own profit alone. It will be a mistake, however, for executives to be so "public-minded" in the promotion of economic growth that they forget an even more important role—that is, as leaders responsible for seeing that people within their enterprises work hard, with satisfaction, and willingly. To do this, it will be necessary for them to develop a genuinely democratic leadership.

It is useless to enter here into an academic argument on

the meaning of "democracy" or "democratization." Some people would say that the word has been used so randomly in postwar Japan that it has come to be an empty prayer, or an ornament. I do not deny that there is much confusion in the usage of the term. "Democracy" has not become an empty word, however. Nor is the quest for it meaningless. Despite common misuses of the term, there is a definite criterion by which a social relationship can be assessed as "democratic."

One is likely to associate "democracy" with "human rights," "equal opportunity," "fair treatment," and the like. The degree of democracy can also be measured, however, in terms of the extent to which members of a group, irrespective of their statuses, are able and willing to participate in decisions of primary concern for the group and its members.

In any group, the essentials for leadership are maximally to fulfill the following two conditions: that each member of the group should be able to satisfy his needs and desires and to develop his talents and capacities; and that the group as a whole should be able to operate as efficiently as possible, to attain its goals, and to perform its social functions. It is not easy to fulfill both conditions at one and the same time, for they usually pull in opposite directions; nevertheless, both are the tasks of leadership.

In a democratic group, these two duties of leadership are joined by a third, namely: that each member be allowed to participate in managerial decisions. In an industrial setting, these requirements are to be satisfied as follows:

1. First of all, employees should be treated as partners of management, not merely as its dependents or as a labor force. This kind of employee treatment contrasts sharply with paternalism, which often regarded employees as "our servants" or "people depending upon us like children."

2. Employees, as partners, should be given training of a

job-centered type. By "job-centered" I mean that training should be aimed at imparting not only the skills and knowledge necessary within a particular enterprise, but also those needed for the same job anywhere else. This is quite distinct from the type of training in "company folkways" which was common under paternalistic management. Some executives may protest that one of the most important but difficult tasks now faced by management in Japan is to develop workers who are loyal to their company, and that job-centered training will lead to the disappearance of company loyalty, which is already everywhere declining. In rebuttal, it should be noted that job-centered training will produce more employees who are primarily work-oriented. Moreover, what if workers, given such a training, also have a deep attachment to their company and are proud to be members of it? Is it not precisely the task of management today to achieve this end? It will in fact be difficult for an enterprise to expect loyalty from the majority of its employees, so long as it is not attractive enough for them to stay, even if management endeavors to detain them by giving them the "company folkways" type of training.

3. Given job-centered training, each employee should be placed in the job most suited to him, where he has plenty of chance to develop his own competence through hard work. He should then be rated in terms of accomplishment, and rewarded accordingly, in contrast to the traditional system of rewards based primarily on length of service. Moreover, employees with special talents and of outstanding performance should be singled out for promotion, regardless of educational background, age, or seniority.

4. More important, employees should be given ample opportunities, and even encouraged, to participate in managerial decisions. This is a logical outcome of the treatment of employees as partners. Inasmuch as the actual procedures

to be followed in promoting worker participation in management will be discussed in more detail in Chapter Six,[11] it will suffice here to cite only a few examples. At the shop level, employees may be encouraged to attend periodic shop-level conferences and freely to discuss means for carrying out the production program of the workshop most efficiently. The next step may be to induce them to take part in the actual process of devising a more efficient and, from their viewpoint, a more satisfactory method to attain production goals. At the level of the enterprise as a whole, it is to be recommended that employee representatives from time to time meet with a matched group of company representatives to discuss basic policies, and to take part in managerial decisions relating to all aspects of the operation of the firm. The suggestion system can also be a good device for promoting workers' participation, when operated in connection with the results of an employee opinion survey. Based upon the data obtained from the survey, employees can discuss and criticize current company policies and, through group suggestion, can influence the managerial decisions necessary for the reform of policy making.

5. Finally, each employee should be responsible for observing the rules and carrying out the tasks which have been decided upon with his participation. Observing rules and disciplines is a constituent part of industrial democracy. By contrast, connivance and favoritism were customarily considered to be the "humane" policy under paternalistic management.

A few more words, in this connection, should be added about workers' participation in management. First, "participation" is often highly touted by industrial psychologists as an effective incentive for employee morale. In actuality, it is more than an incentive device. Workers' participation is the

11. See Chapter Six, section on "Institutionalizing Workers' Participation."

logical result of the premise that workers be treated as partners of management. Being partners, they have the *right* to participate in managerial decisions. Realization of this right is the essential prerequisite for democratizing management.

Second, "promoting participation" does not mean merely to persuade workers to imagine that they are taking part. In its proper sense, it means giving them a real chance to participate and encouraging them to do so spontaneously. An imaginary participation can at best heighten worker morale only temporarily, and a managerial policy deliberately aimed at this result would be a manipulative, not a genuinely democratic, type of management.

Third, by "workers' participation" I mean primarily participation by the employees actually working at an enterprise. I do not mean participation by the labor union, though I know there are a number of cases where unions have attempted to take part in managerial decisions. To my way of thinking, the social role of a union is to function as an "opponent" of management. From the beginning of trade unionism, there have been conflicts of interest between unions, whose primary concern is to protect the lives and rights of workers, and management, whose central aim is to raise the productivity and profits of the enterprise. Through negotiation, generally called "collective bargaining," both parties try to adjust their interests. Union activities vis-à-vis management, as a general rule, center around opposing, restricting, or criticizing managerial decisions. It is beyond the scope of a union's social role to take part directly in the making of managerial policy, to cooperate with management in the carrying out of this policy, and, in consequence, to assume the responsibility for the operation of the firm. In contrast, it is the nature of employees to cooperate with management, and because of this role, they have the right to participate in managerial decisions.

Rationalization versus Democratization

Democratization of industrial management is an urgent necessity in Japan today, because there are nascent but inevitable trends toward democracy both within and outside the enterprise. Democratization alone can meet the needs of the times.

Why, then, do business executives hesitate to take this action? There is no doubt a variety of reasons—subjective and objective, false and unavoidable. In the following, only a few factors which appear to prevent executives from launching upon this course will be pointed out.

For one thing, there is still a good deal of nostalgia among executives for "the good old days" of paternalistic management. Self-justifying arguments, resulting from the "East is East and West is West" type of theory, help to strengthen this sentiment. Warmed by the theory, some executives think that they can get along indefinitely even with a multitude of premodern and undemocratic practices.

Again, there are of course a number of industrialists who are too cautious or too indecisive to choose a clear path at this period of transition.

There is still a third group who wants to avoid the movement toward industrial democracy chiefly because of their misunderstanding of its meaning. Some, for instance, confuse democracy with *laissez-faire* and think that democracy will necessarily make them indulge younger workers. On the other hand, a few are apt to confuse employee participation with union participation, and so they blindly reject the movement toward democracy, fearing that union participation will lead to an infringement of their managerial prerogatives and to workshop strife.

18

One of the obstructive factors in this respect, however, is a managerial policy which I call "manipulative" but which many executives erroneously believe to be the democratic one. For example, they think it democratic for the head of a workshop to assume a superficially friendly attitude to rank-and-file workers—approaching them with a smile, calling them by their first names, or chatting with them about their families. This type of policy, which may be called a "smile and pat on the back" policy, is an important genre of manipulative management. Another technique management uses to give workers an only imaginary or false sense of participation is to shower them with newsletters, white papers on management, intra-company broadcasts, and a variety of other information specially prepared for this purpose. The "human relations" approach, as commonly understood and employed, an approach which has been popular in the recent past, is also a kind of manipulative technique.

There is no doubt that manipulative approaches are aimed at mitigating the workers' latent or manifest resistance to management, and heightening their work morale. Though seemingly democratic, this type of personnel administration actually is far from what I have in mind. These are approaches by which management tries to save the time and expense that will be needed to install a genuine democratic leadership, and so are nothing but a cheap substitute. They attempt to hypnotize and deceive the workers, and so violate the spirit of democracy. For these reasons, they cannot guarantee high morale and motivation among workers over the long run.

More obstructive to the movement toward democracy, however, is the fact that many executives seem still to be preoccupied with the belief that industrial rationalization is more important than industrial democracy. "Rationalization" here refers to any measure to increase efficiency, by introducing more mechanization and automation of equipment and production processes, as well as by a further bureaucratization of the organizational structure. According to this point of view, a fully rationalized plant will produce so efficiently that company goals will automatically be attained, regardless of morale and motivation among the workers. There will certainly be no need to rush headlong into a democratic leadership, the major purpose of which is "only" to satisfy and motivate the workers.

Indispensable as it is, rationalization is not everything. The notion that once an enterprise is fully rationalized everything will go well is decidedly wrong and closely related to the logic of the career technician who considers that workers are nothing but "tenders of machines."

The fatal drawback of this type of thinking is that it overlooks an important fact: rationalization of industry will never be successful unless it is supplemented by democratization. For one thing, rationalization inevitably introduces changes into an enterprise—changes not only in equipment, machinery, and the process of production, but also in the selection, training, placing, and rewarding of workers. Any decision for change, if introduced without due participation of the workers, will cause them to resist it, and eventually the management which is responsible for it. Resistance may not take the form of direct action, but it will lower work morale and decrease work efficiency. Furthermore, a rationalized work environment tends to dehumanize the workers, an effect that in its turn leads to a decline in worker morale and productivity. Rationalization will indeed be counterproductive, unless it is offset by measures that democratize managerial practices.

Executives may try to rely on the traditional methods of paternalistic management to counteract the decline in morale produced by rationalization. This attempt, however, will be in vain, since these methods have lost their once magical power to reduce dissatisfaction and resistance among the employees. Baffled by this result, executives may adopt a manipulative approach. However, the result will be the same or, because the innate deception of this approach will sooner or later create a revulsion among the employees, even worse. Some worldly-wise executives say of the younger workers: "They naturally tend to resist. As they get older and serve longer in the company, however, their resistance will weaken. Until then, all we have to do is to give them more time to enjoy their leisure, through which they can get rid of their complaints and dissatisfactions." I disagree. The complaints and dissatisfactions workers have in their work place will never be dissipated in this way. Only through a democratic leadership can management expect workers to keep their morale high and their motivation to work strong.

CHAPTER TWO

Technological Innovation
and Human Problems

What Are the Human Problems in Industry?

It is often said that technological innovation, and automation in particular, will "liberate man from labor." This hopeful expression is of course not entirely true; innovation will not eliminate the need for human labor in industry. Apart from the fact that we have not yet arrived at the age of the "unmanned factory," probably no enterprise can ever be operated without some human work, no matter how advanced technology may be. Certainly, at the current stage of technology, human abilities, and especially scientific and technical abilities, are as necessary as ever, and there is no doubt that this situation will continue for some time to come.

In Japan, in fact, changing technology has, particularly since the late 1950s, been accompanied by a growing sentiment for the need to develop human capacities still further. The need for talent has become even greater, resulting in the fad of a new catchword, *ningen kaihatsu* (human development). Why, then, in this situation, have human problems become a matter of concern, and perhaps especially in those industries where technological change has gone the furthest? What exactly are the human problems in industry?

Briefly, what is at issue here is the confrontation of man and the machine, or in other words, the contradiction between incessant progress in technology and equipment and the growing demand for freedom and self-realization of workers. Since the industrial revolution, this contradiction has been a striking accompaniment to the modernization of the production process, and it will be necessary for us to consider how industrialization affects workers.

Why, then, have human problems accompanied industrialization? Probably the most common answer to this question would be: because industrialization, in spite of its valuable contributions to human happiness, tends to deprive workers of joy in, and of the will to, work.

If workers in a majority of industries had been truly interested in their jobs, sufficiently rewarded for their performance, and felt that they were their own bosses, then this sort of problem would never have developed. If, as industrialization progressed, those who felt their work really worth doing had grown in numbers, the problems that arose in the early years of modern industry would not have survived. But the reverse is the case, and there is in fact an increase in the numbers without the will to work. Thus, human problems are becoming more, rather than less, of an issue in industry.

There are of course many reasons why a worker cannot take pleasure in, or find in himself the will to, work. Perhaps he lacks the ability or the training necessary to the task. Perhaps the assigned job is unsuitable for him. Quite possibly, he made a mistake in choosing his occupation or his place of employment. We must also never forget that workers are usually compelled to work under the managerial control of the employer. From the time when a worker is employed in a company, he has to submit himself to the policies and regulations of its management. The submission of oneself to such control must make many workers less well motivated than if they could work freely and independently, as in the premodern handicraft industries.

It is those factors, however, that are related to the technology and the managerial system of modernized industries that have contributed extensively to the workers' frustration and dissatisfaction. Although such factors are numerous, it is clear that one of the most important of them is the rationalization of the production process and of the system of management. Rationalization has been the basic condition for the increased efficiency, mass production, and mass consumption that have resulted from the modernization of industry.

Three elements are involved in the process of rationalization, namely: mechanization and automation of equipment and work method; specialization, simplification, and standardization of the worker's job; and bureaucratization in the system of management.

Of these three, the first two hardly need explanation. By "system of management" I refer to those institutions and practices concerning the division of labor and the relationship among the various departments of a firm, such as manufacturing, accounting, sales, labor relations, and so on. Included also are the division of duties within a department as well as the differentiation of status and role among a department's employees. The rationalization of the managerial system, which generally accompanies the growth in size of an organization, is aimed at increasing the efficiency of the firm as a whole, by clarifying the rights and responsibilities of each person, and by restricting the subjectivity of his judgments and the arbitrariness of his conduct. The degree of rationalization of any organization may in fact be measured by the extent to which the kind of rational control in the managerial system has been diffused throughout the organization.

However, where the level of rationalization exceeds certain limits, it will in turn lead to deterioration in human relations among those in different levels of an enterprise, and will increasingly hinder employees from exerting their full talents and abilities. Moreover, there will be a growing tendency to limit the right to make managerial decisions to those in the top echelons of the enterprise. Excessive rationalization of this sort often occurs in large-scale enterprises utilizing highly mechanized equipment. This phenomenon can be called "bureaucratization" in the managerial system.

The aim of rationalization, as was stated above, is to increase industrial efficiency and productivity. After the industrial revolution, rationalization to these ends first spread through Western Europe and North America, and then rapidly developed in Japan and the Soviet Union. Subsequent to World War II, it has spread among the less developed countries of the Near East, Southeast Asia, Latin America, and so on. Needless to say, rationalization has developed even more rapidly in recent years, with the general postwar advances in technology.

It is undeniable that rationalization has had a number of benefits for workers. First of all, it has improved their work environment and eased their toil. Machines have taken their backbreaking tasks, and standardization of their jobs has relieved them of the necessity to exercise ingenuity in their work. Simplification has allowed workers with less education and little training still to perform a useful task.

These points are probably best exemplified by assembly-line operations using belt conveyers. A belt conveyer allots a worker extremely simple tasks, like putting together a set of tiny parts all day long. Since this kind of work does not demand strenuous effort, even a woman or a child can do it.

The job also requires little responsibility; all that is needed is to repeat day after day the process of assembling parts laid out before the worker. Moreover, one result is certainly to increase the efficiency of workers, to raise their salaries, and to improve their working hours and welfare facilities.

There are even wider benefits of rationalization. Mass production results in a reduction of the cost of consumer goods and durables, so that goods which were once beyond the grasp of workers are now relatively easily obtained by them, owing chiefly to mass production by the workers themselves. At the same time, thanks to shorter working hours and the diffusion of facilities for mass recreation, they now have more opportunities for enjoying their leisure. With improvements in mass media, they are also coming to be more and more aware of political and economic events taking place in the wider world, outside the factory.

It is questionable, however, whether the real benefits of the current stage of industrialization have, all things considered, improved the conditions of workers, including white-collar workers. Even today, of course, and particularly in smaller enterprises, many are working under antiquated, even miserable, conditions not unlike those of the preindustrial stage. The major point at issue here, however, is that rationalization has, even in large-scale firms, produced a variety of undesirable effects upon workers. The mechanization and simplification which have lightened their physical burdens, and the bureaucratization of management which has relieved them of all but routine responsibilities, have also robbed workers of self-fulfillment and of spontaneous enthusiasm for work.

It is not necessary here to replicate accounts of the evil influences of industrialization given by Robert Owen, Karl Marx, and other thinkers. Rationalization has indeed turned workers into mere tenders of the machine. This is, moreover, a thoroughly logical outcome of rationalization, since its aim is to replace the human elements of production by mechanization and automation. In this sense, rationalization has always been the process of "dehumanization." We should also realize that this is equally true of bureaucratization of the organizational structure. Those on the bottom levels of a bureaucratized enterprise tend to feel themselves little more than tiny cogs in a mechanism, confined to an insignificant part of a giant machine. It is only natural that they lose their will to work.

It is precisely as this kind of situation has become chronic that human problems in industry have become a conscious issue. The damage from this condition is visited, however, not only on workers, for as their motivation weakens, their productivity is not likely to remain at a high level. The increasing number of such workers thus presents management with a serious paradox, since the purpose of rationalization in its view has been to increase productivity and attain prosperity. This has particularly been management's motivation in postwar Japan where, after the defeat and destruction of World War II, these steps were vital if the country were to regain its position among industrialized nations.

The Impact of Technological Innovation on Human Labor

The term "technological innovation" has been commonly used in Japan since it first appeared in an *Economic White Paper* published in 1956 by the Economic Planning Agency.[1] In Western industrial countries, the term usually does not apply to advanced forms of mechanized production other than automation, but in Japan it is used to cover various forms of advanced technology prior to that stage, including the conveyer system of production.

1. Keizai Kikaku-cho (Economic Planning Agency), *Keizai hakusho: Nihon keizai no seicho to kindaika* (Economic white paper: The growth and modernization of the Japanese national economy; Tokyo: Shiseido, 1956), p. 33.

The word "automation," on the other hand, often refers to one of the following three types of production system. First, a system consisting of a variety of machine tools which are so combined that they function continuously. Transfer machines carry the products in process from one set of tools to another. This type, used widely at automobile assembly plants, television and washing machine factories, and the like, is called "mechanical automation."

Second, a more advanced form of operating system equipped with a feedback mechanism for controlling temperature, pressure, speed, and flux required for the production. This type, introduced mainly in the continuous-process industries, such as oil refineries, chemical works, steel manufacturing and electric power plants, is called "process automation."

There is a third type which is a system of mechanized or automated office work, designed to increase efficiency in recording, adjusting, and producing information by the use of computers, calculating machines, teletypes, and other electric or electronic business machines. This type, used widely in all kinds of office work, and particularly in finance, insurance, and communication business, is commonly called "business automation."

As was stated earlier, technological innovation is a rationalization of the production system, which invariably brings some changes into an enterprise. Though they immediately are changes in technology, equipment, or the work process, they are usually accompanied by a variety of indirect alterations that affect human factors within industry. These may be grouped under the following four types of change: in work force composition; in job types; in workshop structure; and in working conditions.

By "change in work force composition" I mean structural changes in numbers, age, educational career, and so forth, of the work force in an enterprise, as well as in the industry of a country as a whole. Advances in technology, as a rule, bring about a general reduction of the number of workers necessary for an enterprise. As a result, if a country is experiencing depressed conditions, and the growth of business is stagnant, such a change may cause a general increase in unemployment. In the case of Japanese industry, however, as the national economy has continuously expanded ever since technological innovation began to spread, no wholesale discharge of workers, creating a considerable amount of unemployment, has ever occurred.[2] Rather, because of business expansion, most automated industries have suffered from a shortage of labor.

The labor shortage, and particularly that of skilled younger workers, has been a serious problem in Japan since the beginning of the 1960s. According to a *White Paper on Automation* published in 1962 by the Ministry of International Trade and Industry, the total number of skilled younger workers engaged in monitoring and measuring jobs in automated process industries doubled during the three years from 1959 to 1962. The number of skilled younger workers occupied in automated office work also increased for the whole industry 2.04 times during these three years.[3] Again, labor white papers inform us that the shortage of skilled workers in general for all manufacturing industries amounted to 910,000, or 21 per cent of the total number of skilled workers then employed, in 1961, and 1,320,000, or 19 per cent of the total, in 1968.[4]

In contrast to the growing demand for skilled younger workers, older employees who fail to keep up with the speed

2. The rates of unemployment for the whole industry in Japan were 1.8 per cent in 1955, 0.8 per cent in 1965, and 1.2 per cent in 1968. See Rodo-sho (Ministry of Labor), *Rodo hakusho* (Labor white paper; Tokyo: Rodo-sho, 1969).

3. Tsusho Sangyo-sho (Ministry of International Trade and Industry), *Waga kuni sangyo no automation no genjo to shorai* (The present state and future prospects of automation in Japanese industry; Tokyo: Tsusho Sangyo-sho, 1962), p. 16.

4. Ministry of Labor, *Labor White Paper*, 1962 and 1969.

of rationalized production tend to be in a relatively weak position. On the other hand, technological innovation often initiates atomization and simplification of jobs, with the result that it produces a mass of low-skilled employees suited only for simple, repetitive labor. These changes in work force composition give birth to a new status differentiation among workers. Differences in prestige and the treatment workers receive from the company are often accompanied by strained relations between younger and older employees, as well as between skilled workers who are entrusted with the more respected jobs for controlling production process themselves, and less skilled employees who are assigned simple and routine, machine-tending work.

Advanced technology also gives rise to a change in the type of job occupation of workers. This may be seen, above all, in the change of their work environment. For instance, while old-type electric power plants required a number of boiler men who worked like stokers in steam locomotives, new power plants have replaced them with a small team of skilled workers who run the whole plant by remote control, monitoring the gauges on instrument panel boards and manipulating switches, in a clean, air-conditioned central control room. Again, in old-fashioned rolling mills, men were required to toil and drudge in terrible heat to produce thin steel boards. By contrast, in modern strip mills, a small team of neatly dressed young workers is engaged, inside a large, bright room, in turning out the same products, operating a number of big rollers by remote control.

Together with such changes in work environment, many alterations have taken place in type of occupation, new kinds of jobs having emerged one after another to replace obsolete ones. For example, the introduction of transfer machines has resulted in a decline of such jobs as porters and carriers; new electric welding techniques have abolished the jobs of riveters in ship building; and the popular use of small-sized computers has diminished the meaning of skills in the abacus. Besides, automation has created a great many new jobs that call for advanced technological knowledge, such as those of skilled machine repairers, equipment installers, and automated operation controllers. Business automation has also produced a number of new white-collar jobs, such as planners, programmers, key punchers, teletypists, and computer operators.

As a result of the alteration of old and new types of jobs, traditional skills depending primarily on intuition, knack, or manual dexterity have gradually become useless. In their place, a new type of competence comprising advanced technical knowledge, cool and precise judgment, ability to make appropriate decisions, and a high sense of responsibility has come to be regarded as important. At the same time, the traditional status system in a workshop and the enterprise as a whole, which was based primarily on the seniority and educational background, is gradually being replaced by a new one, under which those who are well equipped with the new type of competence will be better treated, regardless of their age and length of service.

Changing technology has also had its effect upon human relations in the work place. As is typical in chemical plants and oil refineries, a workshop increasingly tends to consist of a much smaller number of workers than before, a situation that creates a closer teamwork among its members, though at the same time it gives them a stronger feeling of loneliness.

On the other hand, a workshop composed of a small number of more highly educated and better qualified workers causes a change in hierarchical relations. In Japan, for example, there were traditionally several grades among ordinary workers, and above them were three or four ranks of supervisory personnel, such as the squad leader, assistant foreman, and foreman. In many automated firms today, however, there are only three or four ranks altogether, including general operators, the group leader, and the chief operator. Such a change naturally brings about a shortening of social and psychological distance between shop-floor operators and their superiors, and, as a consequence, workers have more chance than before for personal contact with supervisors, engineers, and other technical specialists.

Furthermore, jobs under automation give workers a better understanding of the whole process of production, and of the close job relationships between their own and other workshops. At the same time, they are now aware, with a growing sense of responsibility, that any blunder or delay at their own posts will affect the flow of production of the whole factory.

It should be noted, on the other hand, that technological innovation in many cases causes transfers of posts. Wherever new jobs are created by an automated operating system, there must be a transfer of existing personnel and the recruitment of new employees. The diminishing number of workers, as well as the simplification of hierarchical relations in a workshop, are also made possible by transferring employees from one post to another within a company. The transfer of posts, though inevitable for the establishment as a whole, can be quite annoying to each employee concerned. It means a reshuffle of a work group, and it causes, even if temporarily, a disturbance in the existing order and security of its human relations. The resulting confusion in work groups is likely to cause the morale and efficiency of the employees concerned to deteriorate.

Finally, technological innovation usually means a change in the working conditions of employees, for it increases per capita output, which in its turn strengthens their demand for better working conditions. Moreover, younger workers today are no longer satisfied with the traditional system of wage and promotion based mainly on age and seniority. A new system must be contrived, in which the content of the job, the ability needed for it, and the degree of work efficiency actually resulting from it are to be the major indices for grading employees.

At the same time, as part of the improvement of working conditions occasioned by the increased productivity of an automated operation, a reduction of working hours is likely to be demanded, mostly through the labor unions. In my opinion, shorter working hours are a prerequisite not only for individual workers to enjoy their leisure, but is, and will be more in future, an important means by which the society as a whole prevents a part of the work force being made superfluous by automation.

The Bright and Dark Sides of Automation

The foregoing brief survey of the changes which advanced technology brings about will show that there are both bright and dark sides of its effects upon workers.

If we try to confine ourselves to the bright, or positive, aspect of its effects, we may enumerate the following points: (1) technological innovation, and automation in particular, will liberate workers from monotonous, repetitive labor; (2) it will enable workers to regain their freedom by bringing the machine under their control; (3) it will make the workers' jobs more meaningful by giving them a view of the whole production process; (4) it will diminish physical fatigue; (5) it will provide workers with increasing chances to exert their abilities, because of the growing demand for special talents; (6) it will enable workers to have pride in their jobs as a result of the heavier responsibility they will assume for their work; (7) it will strengthen teamwork within the workshop; (8) it will shorten the social distance between workers and superiors, so that they will have more personal contact with one another; and (9) it will replace the traditional system of wage and promotion by a more reasonable one based primarily on the workers' real abilities.

If we consider such positive points alone, the prospect seems so bright that the human problems discussed earlier—namely, the general decline of the workers' spontaneous will to work—will gradually be resolved. In other words, as automation progresses, there will be a growing tendency for workers to be freed from the frustration and self-estrangement they used to have in the work place, and instead to feel that they are their own bosses.

In fact, there are a good many advocates of automation who hold such an optimistic view. They are found not only among those managers and technical advisers who have introduced advanced technology in their plants. Among men of learning also, it is by no means rare to find those of similar opinion.

According to the French sociologist Alain Touraine, for instance, three stages can be distinguished in the evolution of manual work. At the first stage, characterized by the skilled manual work of craft production, the principal form of operation was for work to be completed by a small team of skilled workers, with a master craftsman as their leader. Although a variety of tools and machines were used even at this stage, their function always remained subordinate to human labor. At the second stage, characterized by assembly-line mass production, on the other hand, routine, less skilled manual operations, connected with each other by a belt conveyer, became pervasive. At this stage, which lasted until quite recently, machines and equipment were regarded as more important than human factors, and the adjustment of the human factor to the machine was the major concern. It is chiefly in such a monotonous, repetitive type of operation that workers suffer from frustration and alienation. At the third, and more recent, stage, represented by fully automated production, however, operations such as monitoring and measuring the gauges on instrument boards, or manipulating automatic equipment by remote control, come to be predominant, with the result that man gradually recovers his original role of controlling the machine. At the same time, by virtue of the aforementioned beneficial effects of automation, workers come to regain a genuine enthusiasm for work.[5]

Alongside the bright view, however, there is also a pessimistic one of the effects of automation, which is held by no fewer people. They argue that technological innovation is after all a process of intensified rationalization of the production system and that it will inevitably result in an increase in monotonous, dehumanized, and meaningless manual labor, or at least that it will never diminish this type of work.

This argument, in its turn, seems to be based on the following points, which are considered to be the negative effects of advanced technology: (1) technological innovation, and particularly automation, will increase monotonous, repetitive manual labor; (2) it will deprive workers of their freedom by making the machine set the work pace; (3) it will dehumanize jobs by turning workers into mere tenders of the machine; (4) it will intensify psychological tensions; (5) it will reduce the chance for workers to exert their abilities, because of the increasing simplification and standardization of labor; (6) it will increase impersonal organizational control over workers as a result of the automated work process; (7) it will make workers feel more lonely and isolated in the work place; (8) it will intensify interpersonal tensions by introducing discrimination among varied types of workers; and (9) it will create an incessant anxiety about personnel cuts, owing to the increased work efficiency of automation.

Besides these phenomena arising within the enterprise,

5. Alain Touraine, *L'évolution du travail ouvrier aux usines Renault* (Paris: Centre National de la Recherche Scientifique, 1955).

technological innovation can produce certain undesirable effects upon society as a whole. Especially, the spread of automation can increase the numbers of the unemployed and underemployed, and enlarge class distinctions and social unrest.[6] If emphasis is placed solely upon the latter points, it is only natural that many people will be pessimistic about the effects of advanced technology.

Which, then, is truer of these contrasting opinions? We should recognize that these two views simply represent the merits and demerits of the same production system. In addition, we can expect that the negative effects will be emphasized by defects in managerial policy of the enterprise where automation is introduced. In any event, it is erroneous to assert that either one of these two aspects alone is closer to the truth. In other words, all the points I have itemized above are more or less possible effects we may expect from technological innovation. What is important, therefore, is that conscious efforts be made to promote the bright sides of automation and to repress its negative effects upon workers.

What Do Workers Think of Automation?

So far I have discussed the effects of technological innovation mainly from a theoretical viewpoint. What, then, do workers themselves actually think of these effects?

According to the results of a survey reported in the *White Paper on Automation* quoted earlier, out of a sample of 428

6. According to the Swedish economist Gunnar Myrdal, it is evident that American society today, despite its apparent affluence and prosperity, shows an increase in structural unemployment, an enlargement of class distinctions, and the creation of an "under-class," as the result of the spread of automation. Gunnar Myrdal, *Challenge to Affluence* (New York: Pantheon Books, 1963), chaps. 2 and 3.

plants studied where process automation had been introduced, 171, or about 40 per cent, had carried out an overall
transfer, or reassignment, of the employees. It is reported that
as many as 88 per cent of these plants had effected the transfer "smoothly," without any resistance by the workers. By
contrast, in another group of 603 plants which had adopted
mechanical, rather than process, automation, 386, or 64 per
cent, had accomplished personnel transfer, and of these only
61 per cent had gone "smoothly." This means that the latter
experienced more resistance by the workers than the plants
where process automation had been introduced.

When a general reaction of the workers to the introduction of automation was asked for, however, it is reported that
in 65 per cent of these factories where mechanical automation had been adopted, a large majority of the employees
"positively supported" its introduction. Again, in the case
of business automation, of 724 establishments studied, the
employees of 80 per cent of them were reported to have
assumed an affirmative attitude, even when automation was
first adopted a couple of years earlier. At the time of the
survey, this proportion had increased to 97 per cent.[7]

There is, however, another aspect of the workers' reactions.
Viewed in the light of our own research, workers do not
necessarily show such a favorable attitude toward automation. From our survey conducted in 1960 at a steelworks near
Yokohama, for example, we obtained the following data.
The plant was partially automated when it was newly built
in the preceding year, and about half of the employees working there had been transferred from an old-type factory belonging to the same company. At the time when they were
told of their transfer, their reactions to the automated operations at the new plant were, on the whole, reluctant. When
asked how they felt at that time, only 53 per cent answered

7. Ministry of International Trade and Industry, *Automation in Japanese Industry,* pp. 50–51, 143–144, 198.

positively, while 16 per cent were distinctly against it. Such reactions had not yet taken a more favorable turn even at the time of the survey. When asked what they now thought of the automated operations, only 56 per cent of the same workers answered favorably, while 11 per cent were unfavorable. Moreover, contrary to what many proponents of automation predict, 70 per cent of the employees felt that their work load had increased, rather than decreased, in the new plant. Again, 48 per cent of them complained of more fatigue at the automated work than at the old-style operations; only 14 per cent felt less fatigue in the new workshop.

How, then, do workers react to different types of automated jobs? To examine this point, we must turn to the results of another survey of ours. In the summer of 1967 a research committee was set up, under the sponsorship of the Japanese Ministry of Labor, to investigate the reactions of industrial workers to "monotonous work" caused by technological innovation. The committee consisted of specialists in a variety of fields, including sociology, economics, psychology, physiology, and hygiene, with myself as chairman. A survey was planned and conducted by the committee in the same summer, taking a sample of 1,113 workers, all engaged in automated or semiautomated jobs, from fourteen plants and offices of seven different firms in the Tokyo–Yokohama area.

Their jobs were classified into three main categories. The first comprised a variety of assembly-line or mechanically automated operations, performed mostly by the use of belt conveyers and transfer machines, such as those of assembling tape recorders, television sets, washing machines, or automobiles. This type of job may be called "assembly-line work." Five plants manufacturing electric appliances, one automobile assembly plant, and two automobile parts assembly factories were selected to represent this type of work.

The second consisted of such mechanized jobs in automated business offices as those of key punchers, teletypists, and computer operators. The jobs of programmers and analysts, however, were not included. This type of job may be called "business-machine work." A large insurance company was chosen to represent it.

The third included those monitoring and controlling jobs performed in plants where continuous-process automation had been introduced. An oil refinery, a strip mill in a steel manufacturing firm, and three newly built electric power plants were picked out to represent this type of job, which may be called "process-control work."

A paper-and-pencil questionnaire containing fifty-one items was administered to the sample. Following are the workers' reactions to the more important questions.

1. About 58 per cent of the respondents, regardless of job category, reported that they usually got tired of work within two hours after starting it, feeling it to be too monotonous. When respondents with different types of job were compared, however, nearly 65 per cent of those engaged in assembly-line work complained of boredom, as against only an average of 28 per cent of those occupied with process-control jobs.

2. Only 18 per cent of assembly-line workers reported that they felt an overall physical fatigue after a day's work, whereas about 35 per cent of process-control workers did. Similarly, only 34 per cent of the former complained of psychological tension connected with the work, while as many as 53 per cent of the latter did. However, even more workers, over 60 per cent of those engaged in business-machine work, complained of nervous tension.

3. An average of 22 per cent of the total respondents replied that they liked their present jobs. This figure is considerably lower than the number of employees in large-scale Japanese industries of all types who answered affirmatively the same question, in a series of employee attitude surveys I

have conducted since 1952.[8] In the latter case, where only a portion of the workers were engaged in automated jobs, an average of 40 per cent reported that they liked their present jobs. On the other hand, there was a marked difference, in the recent Monotonous Work Survey, between the reactions of assembly-line and process-control workers. While only 14 per cent of the former liked their present jobs, as many as 52 per cent of the latter answered in the same manner.

4. When asked if their present jobs enabled them fully to exert their abilities, an average of only 28 per cent of the respondents of automated industries answered affirmatively, again considerably fewer than the approximately 47 per cent of general industrial workers who gave the same answer. When the respondents' attitudes were compared by job type, however, only about 23 per cent of both assembly-line and business-machine workers replied that they could exert their abilities in their present jobs, whereas as many as 59 per cent of process-control workers answered affirmatively.

5. When asked if they wanted to continue in the same jobs, only about 19 per cent of assembly-line workers, as compared with about 45 per cent of process-controllers, replied affirmatively. Those who on the other hand wanted to change jobs were almost double those who wanted to continue in the assembly-line case, but amounted to an average of only 25 per cent of all process-controllers.

6. Contrary to expectations, though process-control workers usually work in small groups in an isolated workshop, they were found less frequently to feel lonely than those engaged in the other two types of job. Whereas, for instance, more than 80 per cent of the keypunchers, who worked together in a large group of about fifty people, complained of loneliness, less than 40 per cent of the operators working in a small team of fewer than six members each within the

8. For details of these attitude surveys, see Chapter Four, section on "Company Allegiance and Union Allegiance."

central control room of a power plant had similar feelings. 7. Finally, when asked what they worked for, more in assembly-line jobs than in process-control operations answered that they worked merely to make a living. Though few in absolute numbers, there were more among those in process-control operations who considered that they worked to contribute to society.[9]

From the foregoing, we may conclude that, though workers engaged in automated or semiautomated jobs tend to be dissatisfied with their work as being "monotonous," there are marked differences in the reactions of assembly-line and business-machine operators on the one hand and process-control workers on the other. Boredom and dislike of the work are much more frequently found among the former than the latter. If technological innovation does indeed bring monotony into work, it is only to the former types of job that it does so. Process-control jobs, on the other hand, are not necessarily monotonous, nor are they meaningless for the workers. In fact, as the surveys show, such workers are more satisfied with their jobs than those engaged in ordinary, nonautomated operations. It is to be noted, on the other hand, that process-control work is likely to cause workers more physical fatigue, as well as nervous tension, than the assembly-line does.

A similar conclusion has been reached recently by the American sociologist Blauner, based upon a variety of empirical studies made by himself and by others on the jobs of printers, textile workers, automobile assembly workers, and chemical operators. By analyzing worker attitudes toward different types of job, he compared the latter in terms of the degree of worker alienation and freedom. According to him, there are three major types of blue-collar work: traditional

9. Rodo-sho Tancho Rodo Senmonka Kaigi (Ministry of Labor Committee for the Study of Monotonous Work), "Tancho rodo jittai chosa hokoku" (Report of the [1967] survey of monotonous work; Tokyo: Rodo-sho, 1968), mimeographed.

skilled work associated with craft technology, represented, for example, by the jobs of printers; low-skilled routine operations associated with assembly-line technology, represented, for example, by the jobs of automobile workers; and "non-manual" automated production associated with continuous-process technology, typified by the jobs of chemical operators, in which "responsibility" plays a more important role than manual skill.[10]

Blauner argues that where craft technology dominates, "alienation is at its lowest level and the worker's freedom at a maximum." "Freedom declines and the curve of alienation rises sharply in the period of machine industry," and it "continues upward to its highest point in the assembly-line industries of the twentieth century." With the progress of automated operations by the continuous-process technology, however, "there is a counter-trend, one that we can fortunately expect to become even more important in the future." As automated operations of this type increase, Blauner believes, workers regain their control over the work process. "The alienation curve begins to decline from its previous height as employees in automated industries gain a new dignity from responsibility and a sense of individual function."[11]

It will be obvious that Blauner takes an optimistic view of the effects of automation, similar to that of Alain Touraine cited earlier.[12] Both sociologists believe that monotony and meaninglessness of work, as well as self-estrangement and dehumanization of workers, can be reduced by replacing as far as possible the assembly-line type of manual labor with process-control operations.

Although in principle I agree with their opinion, I cannot do so fully. Even if possible, it will take years, and even

10. Robert Blauner, *Alienation and Freedom: The Factory Worker and His Industry* (Chicago: University of Chicago Press, 1964), pp. 167–169.
11. *Ibid.*, p. 182.
12. Blauner himself admits this similarity. See *ibid.*, p. 169, footnote.

generations, before we can completely replace all existing assembly-line and business-machine operations by the process-control type of work under a more advanced technology. Moreover, mere alteration and improvement of technology and the nature of the work process will not necessarily enable workers to regain their freedom and to be relieved from self-estrangement at work. Here are some empirical data to attest to the latter point.

The aforementioned research committee, set up by the Japanese Ministry of Labor, conducted another Survey on Monotonous Work in the summer of 1968. This time a larger sample of 2,451 employees, all occupied in automated or semiautomated operations, was drawn from fifty-three plants belonging to twelve different industries all over the country. For comparison's sake, an additional sample of 1,387 employees working in conventional, nonautomated jobs was taken from the same plants. A questionnaire similar to that of the 1967 survey was administered to both samples. It was found that, contrary to the results of the previous survey, there were no significant differences in reactions between assembly-line and process-control workers. True, more respondents on the average were found in assembly-line than in process-control work who complained of boredom and meaninglessness in their jobs. However, the differences in percentages of the respondents in the two types of job who had such complaints were at most 8 per cent, and in most cases less than 3 per cent. Though about 20 per cent more of the respondents in business-machine work felt boredom, as compared to those in process-control operations, the percentages of workers who disliked their present jobs were much the same for both types of work. Moreover, in some firms and plants more workers in assembly-line operations felt less boredom or meaninglessness than did those in nonautomated, conventional types of job; conversely, in some other

43

establishments more workers in process-control jobs than in nonautomated operations felt alienated.[13]

Judging from the above, it is hardly conceivable that the problems of "monotonous work" will be successfully solved solely by improving technology and the nature of the work process. In order to restrict the negative effects of automation upon workers and to promote as far as possible what we have called its bright side, it will be necessary to devise some other means, in addition to technological improvement.

What Is the Solution to the Problem?

What measures should be taken to minimize the negative effects brought about by technological innovation upon workers and, at the same time, to enable them to recapture their spontaneous will to work and to have the joy of actualizing themselves in their daily life? What is the principal key to the solution of the human problems in industry?

Generally speaking, there are at least five different approaches, as follows: (1) improving employee welfare and human relations in the work place; (2) facilitating self-fulfillment of workers in their leisure activities; (3) replacing conventional types of production by a more advanced, fully automated technology; (4) promoting job enlargement for workers; and (5) encouraging workers' participation and self-government in industry.

The first approach, which may be called "paternalistic" in a broad sense, consists of such attempts as amplifying safety and sanitary facilities, devising the means to reduce work loads, improving welfare facilities, or enlarging cultural and

13. Ministry of Labor Committee for the Study of Monotonous Work, "Tancho rodo jittai chosa hokoku" (Report of the [1968] survey of monotonous work; Tokyo: Rodo-sho, 1969), mimeographed.

recreational services within the enterprise. The improvement of human relations in industry, represented, for instance, by the encouragement of two-way communications and mutual understanding in the work place, may also be included here. Such devices, though the extent to which emphasis is placed on them may be different, have long been adopted by management in most industrial countries. In Japan particularly, management traditionally attached importance to the paternalistic approach until the end of World War II.

Though indispensable for keeping up worker morale, improved welfare facilities and human relations are after all what the industrial psychologist Frederick Herzberg calls "hygiene factors" for worker motivation.[14] Once they have become established as standard facilities or procedures, their effects invariably diminish as workers get used to them. Without introducing technological improvement or organizational change into the enterprise, we cannot expect these measures to produce any long-lasting incentives for the workers.

The second approach, which tries to minimize the alienation and frustration of workers by helping them find self-fulfillment in their leisure, is a relatively new one. In Japan, this device has come to be emphasized by management since the late 1950s, when the workers' standard of living began to improve considerably, and their working hours gradually shortened. What characterizes this attempt is that it virtually gives up the idea of eliminating workers' frustration and self-estrangement in work and production, on the assumption that the evil effects upon workers caused by technological innovation are after all inevitable. It proposes instead to let the workers have more chance to realize themselves in their leisure by providing them with better wages and shorter working hours, so that they can compensate, so to speak, for

14. Frederick Herzberg et al., *The Motivation to Work,* 2nd ed. (New York: John Wiley, 1959), p. 157.

the damage they suffer in the work place by their free activity in their leisure. Without doubt, such an attempt will provide much benefit to workers in their daily life. In many industrial countries, including Japan, management has recently begun to concede to workers at least two paid days off a week, in the hope that it will improve their morale and eventually raise work efficiency.

The major defect of this approach is that it is no solution at all of what we have conceived as the human problems in industry. It cannot diminish workers' alienation and frustration in the work place, because from the beginning it abandons hope of diminishing them. Proponents of this device argue that in the near future the hours people can spend a week for pastime and hobbies will on the average become considerably greater than those they work. However, for some time to come, and for the great majority, work will remain the most important single life activity, in terms of time and energy. Moreover, workers' dissatisfaction and their sense of powerlessness at work are always projected onto their private life and their leisure activities. If a worker is alienated and unhappy at work, he will quite likely be dissatisfied and frustrated with his leisure, or else he may make a desperate effort to escape from, or forget about, work. He will, at any rate, be unable to make a free and creative use of his leisure.

The third approach may seem at first to be paradoxical, since it attempts to remove the evils resulting from technological innovation by means of promoting a still more advanced technology. Actually, however, what it aims at is to replace the negative effects of machine-tending, assembly-line types of operations, to be found mostly in the semi-automated stage of technology, with the process-control type of work of a more advanced technology of automation. It is proposed that in doing so most of the low-skilled workers, who are now occupied with monotonous, repetitive, and routine labor, will be assigned more complex, challenging, and meaningful jobs. We were well aware, even before Charlie Chaplin, that assembly-line operations using belt conveyers often cause monotony and meaninglessness in work. As a consequence, a variety of devices have been invented, and adopted, by worker-minded management. For example, the "round-table system" has been employed in assembly plants in place of conveyer production. As Blauner believes, if the continuous-process technology can be successfully adopted extensively, as it has been in modern chemical plants and oil refineries, much of the monotony and meaninglessness of work, as well as the alienation of workers, may be removed, since workers then can be relatively free from the mechanized operations, and become concerned with the entire process of production, rather than with only a minute segment of it.

It is inconceivable, however, that we can switch all conveyer-operated jobs over to continuous-process automated operations in the foreseeable future. There are a number of industries where the continuous-process type of operation cannot possibly be adopted, in terms of material used and equipment required. Again, most of the less skilled workers who are now engaged in monotonous, repetitive labor are likely to remain in another kind of standardized, routine job, even after fully automated technology has been applied to most of the industries of a country. Moreover, even if we are successful in replacing every assembly-line job by process-control work in every kind of industry, we can hardly expect to eliminate the alienation and frustration of workers, as long as they have to work, under the control of autocratic management, in a dehumanizing atmosphere of bureaucratic organization.

The fourth approach, which advocates job enlargement as a remedy, attempts to diminish the evil effects of excessive subdivision of labor, usually attendant upon advanced technology. Rationalization of the production system tends to divide a complete work process into a number of minute, oversimplified, repetitive tasks, and to allot each of them to an operator. As a result, workers are likely to find their jobs monotonous and meaningless and to feel themselves deprived of independence and self-fulfillment at work. Job enlargement is an antidote against such evils; by it each worker's job is so enlarged that he is to accomplish all, or at least most, of the operations to be performed in manufacturing a product. In assembling a television set, for example, instead of allotting atomized jobs to a number of less skilled operators working at a belt conveyer, single more highly skilled workers, each sitting at a table with a set of the necessary parts and equipment, are assigned the complete assembly of a television set. Under this system, which is often called "one-man production," workers are able to enjoy greater independence in deciding on work methods, and to assume more responsibility for their products. It is widely believed, in Japan as well as in the United States, that job enlargement has the effect of eliminating the frustration and dissatisfaction of workers. A number of well-known American firms, for example, IBM, Sears Roebuck, and Detroit Edison, have introduced the device, and in Japan, Sony is reported to have been successful in adopting job enlargement.

For all its good effects, however, there are limitations to this method. For one thing, workers who can perform one-man production operations are necessarily confined to the highly skilled; less skilled employees will not be allowed to perform this type of task. Even if all employees in a firm were to be qualified for an enlarged job, it is virtually impossible, in terms of costs and technology, to carry on all jobs in a firm with one-man production methods. Besides, there are a good many companies where the one-man system cannot be adopted at all. Furthermore, job enlargement is inevitably accompanied by a reduction of the labor force of a company. Unless the firm is expanding so that it can absorb the employees who became surplus by job enlargement, its adoption is likely to result in a personnel cut.

Finally, the fifth approach attempts to solve the human problems in industry through the encouragement of workers' participation in management. It has long been known at least by some progressive industrialists that a system of participation by employees has the effect of minimizing their work alienation and of fostering their spontaneous enthusiasm for work. It is for this reason that attempts to democratize the organizational structure of enterprises by introducing worker participation have been supported not only by the representatives of labor but also by certain circles in management. Yet, until the end of World War II, the majority of managers either were ignorant of the effect or else, even if aware of it, were too conservative to make up their minds to adopt worker participation. Some executives considered the system an infringement of managerial prerogatives; others had misgivings about the employees' capability for carrying out effective participation.

Since the end of World War II, however, the need and significance of a system of employee participation have gradually come to be recognized in many industrial countries. In Japan, along with a general decline in the power of traditional autocratic-paternalistic management to induce work motivation among employees, the need for a more democratic method has come to be widely recognized. The growing shortage of labor, as well as a relative increase in labor turnover, have also made Japanese management look for a new way to promote employees' identification with the company. A system of employee participation has been considered one of the most fundamental means to this end. In fact, when successfully introduced, the system will provide employees with greater freedom to decide on methods and pace of their work, and with more autonomy in operating and controlling their daily activities in the work place. It is with good reason, therefore, that general support of the system has also been found among workers. According to a finding of the 1968 Survey of Monotonous Work cited above, for example, out of a sample of 2,451 workers engaged in automated or semiautomated jobs, as many as 80.3 per cent reacted favorably to the introduction of a system of employee participation.

It should be noted, however, that there are varied forms and procedures, with correspondingly varied effects, for workers' participation in management. For example, the form in which a limited number of employee representatives take part in the top-level decisions of management naturally differs from the type in which all employees of an enterprise, including rank-and-file operators at the shop level, govern themselves in the operation and control of the daily activities of their workshops. In my opinion, the most basic, and therefore the most important, effect upon workers can be expected only from the type in which all employees govern themselves. It is desirable, however, that, in addition to this, participation by employee representatives in top managerial decisions be also instituted. This is of course not the place to discuss at any length the different methods of participation and their institutionalization into the formal organization of an enterprise. For details of these points, see the last two sections of Chapter Six.

It would be over-optimistic, however, to claim that the negative effects of technological innovation will be wiped out solely by the introduction of a system of worker participation. What I want to assert here is simply that among the aforementioned five approaches, the last is the most fundamental means of solving human problems in industry, and that without reform in this respect neither improvement of technology nor facilitation of leisure activities can possibly be effective. Besides, unlike job enlargement or technological improvement, employee participation need not be confined to highly skilled employees engaged in only certain kinds of industry. Wherever an organizational change is introduced and training of employees in participative practices is secured, the system can be effectively applied to all kinds of workers employed in all types of industry. The key to the solution of human problems in industry, therefore, is to be found in the attempt to promote all four other improvements mentioned above, but with priority being given to the fifth, that is, the encouragement of workers' participation and self-government.

By way of conclusion, a word may be added about the view that the self-estrangement of workers is basically caused by the capitalistic order of society, and that it can never be wholly eliminated unless the fundamental socio-economic system is reformed. Even today, there still are some who support such an oversimplified theory. They argue that the evil effects of automation are unavoidable as long as the basic social and economic system of society remains capitalistic, whereas under a socialistic order, the introduction and spread of automation would invariably be welcomed by workers and would have the effect of heightening work morale and productivity.

If this theory were true, in a capitalistic society the alienation and frustration of workers could never be removed, even if worker participation were encouraged, the production system reformed, welfare facilities improved, and leisure activities of workers expanded. In a socialistic society, on the other hand, workers would always be happy and satisfied and would have a genuine motivation to work, even if they were under a faceless and autocratic bureaucracy, without any improvement in the production system or in welfare facilities. Fortunately, however, those who understand the real psychology of the human being will never be persuaded to believe in such a naive ideology. In fact, in a socialistic country like Soviet Russia, it has been found that the improvement of the working conditions that directly affect workers must always be given serious consideration.

Without doubt, reform of the basic socio-economic order will have at least some indirect effect upon workers. However, the factors which directly and more deeply influence workers' satisfaction and spontaneous enthusiasm for work are to be found among those that I have discussed above.

CHAPTER THREE

The Younger Generation
in the Era of Automation

Leaping the Generations

Peter F. Drucker, on the occasion of his visit to Japan in the summer of 1960, delivered a public lecture in Tokyo entitled "Managing Tomorrow," in which he said that the most remarkable change one could predict for the 1960s would be "the leap of the younger generation." As technological innovation progressed, the managerial strata who led the country's industry would rapidly be replaced. Whether they wished to or not, those over fifty would have to give way to those under forty. According to him, this would be the biggest change to be expected in the 1960s.[1]

This prediction by Drucker seems to have come true in Japan, at least to a certain extent. Everybody knows that in Japan today the average age of workers is declining, and that in those enterprises where automation has been introduced in particular, the percentage of younger workers with an upper or lower secondary school education is getting

1. Peter F. Drucker, *Asu o keieisuru mono* (Managing tomorrow; Tokyo: Jimu Noritsu Kyokai, 1960), pp. 111–129.

higher year by year. Moreover, in some popular companies which have developed rapidly since the war, such as Matsushita Electric, Honda Motor, and IBM Japan, a bold program of selective promotion has been carried out, resulting in section heads in their twenties and department chiefs in their thirties.

Frankly speaking, however, there is some exaggeration in what Drucker said, as is often the case in public speeches given by foreign visitors. His prediction may not have come true exactly as he said, nor should we take it too literally. What he really meant, I suppose, was that, as a result of technological innovation, those who failed to meet the requirements of the new epoch would necessarily drop behind, to be replaced by those who were well qualified for this era of automation, regardless of their age and generation. He also wanted to stress that it would be a task of the utmost importance for the present industrial leaders to encourage those workers who were well qualified for the new era and, throughout the age groups, to pick out the more qualified for promotion.

In fact, the recent progress of technological innovation has had considerable effect in reforming the managerial organization of Japanese industry. As a consequence, there is a tendency for those industrialists who persist in their traditional managerial philosophy and who are afraid of taking risks to introduce necessary reforms to be left behind and to be compelled to retire from responsible positions. This does not mean, however, that the present industrial leaders, whether qualified or not, will be retired at one stroke, or that the younger generation, regardless of their competence and efforts, will suddenly take over their positions. It would not even be desirable for such a drastic alternation of generations to take place, if Japanese industry is to keep up its successful development.

What is really needed today is for industrial leaders to

recognize the actual change of the times, to understand more fully the feelings of the younger generation, to listen to their voice, and to improve their treatment. It will be particularly necessary for them, in every sphere of the enterprise, to give younger workers sufficient training, to search out talented and able persons from among them, and to select them boldly for promotion. This is precisely the way to realize "the leap of the younger generation." There are, of course, less capable people as well as people of talent among both older and younger workers. It would add nothing to the development of industry if the dregs of the older generation were to be replaced by their counterparts in the younger generation. What Drucker said should be regarded as a piece of bitter counsel to the present leaders, rather than as a pretty compliment to younger people.

Two questions are to be considered in this connection, namely: whether the younger generation is now prepared to make the "leap"; and whether the present industrial leaders are ready to give younger workers the chance. In the following pages I shall discuss these two points.

Is There Any Chance for the Young to Make the Leap?

The first question above may be answered by considering the following three points: (1) the existence of an actual chance for young workers to make the leap; (2) the extent to which they are capable of and trained for it; and (3) the degree to which they are motivated for it.

As regards the first point, let us examine, from the results of recent surveys, the rate of factory workers and lower-level white-collar employees whose sons have moved up to managerial occupations. According to our survey of social stratification and social mobility in Japan, conducted in 1955 with a national sample of about 2,000, of some 400 blue-collar fathers, including skilled, semiskilled, and unskilled, only 3 per cent had sons who, at the time of the survey, held a managerial position, such as section head or above, either in public service or in private enterprise. Of those who at that time were in a managerial stratum, on the other hand, 23 per cent had managerial fathers, 5 per cent professional, 7 per cent white-collar, 35 per cent agricultural, and only 14 per cent blue-collar.[2]

Again, the results of a survey made by Hiroshi Mannari and James C. Abegglen show that, of 967 respondents who in 1960 held a top management position, including that of managing director or above, only 1 per cent had blue-collar, and 9 per cent white-collar fathers; therefore, only 10 per cent had working-class fathers. By comparison, 22.5 per cent of the respondents had fathers who were managers in large-scale industries, 11.5 per cent higher civil servants, and 22 per cent owners of small-sized enterprises; that is, a total of 56 per cent had fathers in managerial strata. Those who had agricultural fathers amounted to 24 per cent.[3]

Similar data were obtained from a survey conducted by Kazuo Noda and Yuichi Yamada. Their data show that, of 452 high executives in some 800 major Japanese companies listed on the stock exchange, 64 per cent were university graduates, 26 per cent graduates of higher schools and colleges, and about half of the respondents had an advantaged family background, their fathers being managers, professionals, government officials, and the like. According to the researchers, most of the remaining half of the respondents were also brought up in well-to-do families of merchants or

2. Kunio Odaka, ed., *Shokugyō to kaisō* (Occupations and social stratification; Tokyo: Mainichi Shinbun-sha, 1958), pp. 120–122.

3. Hiroshi Mannari, "Nihon no keieisha no shakai-teki seikaku" (Social characteristics of Japanese management), *Shakaigaku Hyōron*, no. 45 (September 1961), pp. 9–11.

farmers, and less than 10 per cent had only a middle-school education.[4]

Let us compare these figures with corresponding ones for the United States. Research initiated in 1928 by F. W. Taussig and C. S. Joslyn and a similar study conducted in 1952 by W. Lloyd Warner and Abegglen show that, of the top management people studied, the proportions of those who had blue-collar fathers were 11 per cent in 1928 and 15 per cent in 1952, while those who had white-collar fathers amounted to 12 per cent and 19 per cent, respectively. This means that those who had blue- and white-collar fathers totaled 23 per cent of the top managers even in 1928, and as high as 34 per cent in 1952—far exceeding the numbers of similar cases in Japan.

On the other hand, of the same respondents, the proportion who had fathers holding managerial positions in large-scale enterprises or in public service was 31 per cent in 1928 but had decreased to 24 per cent in 1952. From this we may conclude that in recent generations in America, whereas the tendency for managerial people to succeed almost hereditarily to their fathers' occupation has declined, the proportion of managers who come from the working class is increasing.[5]

It is to be noted in this connection that these managers born in working-class families were, unlike their fathers who had a low educational background and had to work their way up, in most cases college graduates. Through their fathers' as well as their own efforts, they enjoyed a higher education and rose rapidly in their careers in large firms. At any rate, we note that, in contrast with the case in Amer-

4. Kazuo Noda and Yuichi Yamada, "Gendai keieisha 452 nin no rireki to sono shakai ishiki" (Careers and social attitudes of 452 business executives today), *Kindai Keiei,* 5 (November 1960), pp. 10–21.

5. W. Lloyd Warner and James C. Abegglen, *Occupational Mobility in American Business and Industry* (Minneapolis: University of Minnesota Press, 1955), pp. 44–46.

ica, the managerial stratum of Japan is still to a considerable extent a closed circle.

So far we have confined ourselves to occupational mobility between two generations. A similar tendency, however, may be found during the lifetime of an individual. According to the first survey cited above, of some 500 respondents whose initial occupations were in blue-collar jobs, only 2 per cent were managers at the time of the survey. When occupational groups are reclassified into three major categories—namely: blue-collar, including farmers; white-collar, comprising clerical and sales; and the "elite," consisting of professional and managerial personnel—then of some 1,000 people of blue-collar origin, as many as 91 per cent still remained in the same category. Only 6 per cent were in white-collar jobs, while only 3 per cent had reached the "elite" stratum. Here again we can see that there are high barriers between the managerial classes and the blue-collar workers.[6]

Such barriers will remain for a while to block the avenues of promotion to managerial positions for the present younger generation. True, the recent shortage of highly skilled young workers, which has been felt most acutely since the early 1960s, seems to have had some effect in lowering the barriers. Moreover, if the progress of technological innovation helps to break down the traditional system of promotion based primarily on *nenko,* or length of service, this will surely make the advance of the younger generation much easier.

Actually, however, the traditional promotion system still persists. As a result, those companies that have recently employed a great many skilled younger workers with at least upper secondary schooling, in order to meet the work requirements resulting from technological innovations, have had to elaborate a new device to rationalize the discrepancy in treatment and promotion between them and the older workers who had only a compulsory level of education. In

6. Odaka, ed., *Occupations and Social Stratification*, pp. 98–101.

fact, they have worked very hard to coordinate a new route of promotion, linked to the degree of technical ability, with the traditional channel based primarily upon seniority.

At the automated factory of a steel manufacturing company near Yokohama, where we conducted an employee opinion survey recently, a "Blue-Sky" promotion system had been publicly announced. It was maintained that by this new system the traditional status discrimination between the shop and the office workers would be removed, and that any talented shop operators, depending upon their effort, could be promoted to the rank of section head or even to that of factory manager. One would imagine that the workers in this factory, relieved from heavy labor under the great heat of the melting furnace and now given the chance to be promoted without discrimination, would have been satisfied with the new working conditions. The results of the survey, however, showed that the reverse was the case.[7]

There must have been several reasons for the workers' dissatisfaction. One of the most important, however, was the fact that, at the time of the survey, the "Blue-Sky" system was still little more than window dressing, because of the difficulty of adjusting it to the old-style promotion system so far prevailing in the other plants of the same company, under which the treatment of older employees transferred to the new factory was determined. Similar examples may be found in other companies where automation has been introduced. We cannot be too optimistic, therefore, about the future prospect that technological innovation will soon break down the traditional system of promotion.

Furthermore, there still remains a strong tendency for Japanese enterprises to attach too much importance to their employees' formal educational background. As a consequence, chances for young workers to make the leap are

7. Ken'ichi Tominaga, "Gijutsu kakushin no naka no rodosha" (Workers under technological innovation), *Economist,* October 11, 1960, pp. 36–41.

still greatly influenced by their school careers. It is to be noted in this connection that, in Japanese companies, workers are often classified into *yoseiko* (trainees), *honko* (regulars), and *shokuin* (staff members), according to their educational background; that is, according to whether they are lower secondary, upper secondary, or college graduates. Moreover, the "rank," or the social reputation, of the school from which a worker graduated often affects the speed and limits of his promotion.

Once a worker has gained a certain post in a firm through his school experience, however, *nenko,* or length of service, then becomes the most important factor in determining his career. Thus, one can roughly estimate, for example, that it will take at least fifteen years for a college graduate, and twenty-five years for an upper secondary graduate, to become a section head. In other words, schooling plus length of service are, in most cases, almost the only measure of a worker's chance to be promoted to a managerial position.

Under such circumstances, it can readily be imagined that there is very little, if any, chance for young workers with a mere lower secondary schooling to make the leap. *The Life of Young Workers,* a report based on the results of a recent survey conducted by the National Federation of Institutes of Educational Research, informs us that a lower secondary graduate generally has to pass the severe competition of an entrance examination, in order to enter a large company. Once he has become a trainee, he then has to work under such harsh conditions that any young man's dream will turn into a nightmare. Even if everything is going well, he will be thirty before he becomes the leader of his work-bench, about thirty-seven when he reaches the position of assistant foreman, and already forty-seven, near the age of his retirement, fifty-five, when he at long last becomes a foreman or a workshop chief and receives treatment as a staff member of the company. Even this career, however, would be rather a lucky one, according to the report. The majority of lower secondary graduates coming from farmers' families, or for other reasons incapable of going onto higher schooling, can at best enter medium- or small-sized enterprises. In such cases, it is almost impossible for a worker, even if he can endure low wages and hard living for a long time, to get to the top echelon of the company unless he has a special connection with its owners. Such circumstances cause most workers to move from company to company.[8]

Viewed in this way, the chance to make the leap seems to come to only a limited number of specially privileged individuals. Even if the younger generation is fully prepared for the leap, there are still too many hurdles—length of service, educational background, fathers' social position, and so forth—for it to succeed. It is even more necessary today, therefore, to remove, or at least to ease, these barriers than it is to concentrate upon preparation of the younger generation itself.

We must not forget, however, that Japan is not alone in finding that people's social origin and educational career influence their rise in occupation. Similar limitations can be found in such European countries as Great Britain, France, or Germany. Even in the United States, where the "American Creed" of equal opportunity has been deeply rooted in people's minds for a long time, education and length of service seem to have now come to be prerequisites for success.

This change is clearly demonstrated, for example, by Vance Packard in his *The Status Seekers.* According to him, there is a growing tendency for American companies to consider college education as one of the requirements for employment. This tendency may be accounted for by management's belief that college graduates are less risky for the

8. Zenkoku Kyoiku Kenkyusho Renmei (National Federation of Institutes of Educational Research), *Kinro seinen no seikatsu* (The life of young workers; Tokyo: Toyokan, 1959), pt. 2, chap. 3.

company. A deeper reason, however, is to be found in the fact that, as large, bureaucratic business organizations have developed, the selection of employees has been standardized and mechanized. Yet, since a college education takes a great deal of money, children from poor families find it hard to acquire it, even if they are talented. As a result, it is becoming more and more difficult for them to obtain a promising post in a large company.[9]

Despite this recent tendency, the barriers between classes do not seem to be so high in the United States as in Japan. As was pointed out above, the number of American boys who were born in workers' families, and who after graduating from a college entered a large firm and eventually acquired a managerial position, increased considerably in the 1950s, as compared with the 1920s. This trend will probably continue for some time in the United States, where college education has been much more prevalent than in most other countries. In Japan, to be sure, as the number of colleges and universities has rapidly increased since the war, the number of working-class children who enter them has multiplied. However, it will probably take more than a decade for our country to be democratized enough for everybody to have opportunties to the same extent as the Americans now have.

Are They Capable of Making the Leap?

The second point noted at the beginning of the previous section is concerned with whether young workers are capable of, and trained for, the leap. Here again the question at issue

9. Vance Packard, *The Status Seekers: An Exploration of Class Behavior in America* (New York: David Mckay, 1959), chaps. 20 and 24.

seems to be that of the particular conditions under which they work, rather than the degree of ability and preparation they have.

In present-day Japan, only a few automated factories give the workers sufficient training for them to adjust themselves to the new techniques and arrangements required by automation. Of those automated plants I have so far visited, most are so eager to recover the enormous investment that they tend to urge the workers to start up a full operation after giving them sporadic or nominal training for only a limited period. In one company I visited, it was explained by management that employees transferred from older plants have a "knack," already acquired, for undertaking any job, no matter how unfamiliar. It was also stated that employees should be able to achieve the expected efficiency, even if they had not received ample training. Perhaps this was true, particularly with Japanese workers, who are known for their dexterity and adaptability.

The question now is to what extent workers are interested in, and satisfied with, automated work, if they are not sufficiently trained. At the automated plant where we conducted research, we found that, despite its well-equipped work environment, nearly half the employees were not satisfied with their present jobs. By contrast, similar surveys recently conducted in the United States, such as, for example, those at an automated factory of U.S. Steel and at a newly equipped power plant in Detroit, show that the degree of satisfaction is generally high, not only among the newly employed but also among those transferred from older plants of the same company. In fact, at the Detroit power plant, over 80 per cent of the employees transferred answered that they "liked" their present jobs better than the previous ones.[10]

10. Floyd C. Mann and L. Richard Hoffman, *Automation and the Worker: A Study of Social Change in Power Plants* (New York: Henry Holt, 1960), pp. 77–79.

Moreover, contrary to our expectation, worker fatigue and the frequency of minor accidents were reported to increase at the plant we studied, rather than to decrease, as is usually the case in American automated factories.

Where do these differences between the two countries come from? In my opinion, one of the major factors is the quality of worker training. While American workers generally are well trained in their new work environment, management in Japan tends to be too much concerned with a quick recovery of the costs of new equipment, so that they are likely to give little consideration to systematic training and the appropriate placement of workers. I suspect that, besides its concern for profit, management often labors under the misconception that workers are no more than tenders of machines, particularly in an automated environment, and for this reason as well neglects worker training.

If even technical training suffers from these circumstances, all the more so does managerial training. True, on-the-job training programs often given to young supervisors and staff members, such as Training Within Industry and Management Training Program, include some instructions on leadership skills. They are concerned, however, mostly with techniques for manipulating their followers, or leadership within a limited sphere of the workshop; they are not aimed at cultivating future industrial leaders to whom an efficient and effective operation of a whole enterprise can be entrusted.

Even more defective and impractical is the training given at schools in this age of technological innovation. As a result of the general tendency to attach too much importance to the formal educational career, most lower and upper secondary schools in Japan concentrate upon a preparatory education, by means of which they can get more students into popular colleges and universities. The vocational guidance and technical training required for getting jobs are given at these schools, but in most cases they are merely extracurricular activities. At colleges and universities, on the other hand, despite the declaration that they aim at developing men of culture who will also be future leaders in many fields of society, the majority of professors give abstract, lofty, but unrealistic lectures. Whereas industrialists may be responsible for the practice of overemphasizing the academic career, it is the colleges and universities, and therefore their teachers, who are to blame for having made education and learning into an impractical pedantry.

A recent report submitted to the Japanese government by the Economic Planning Agency included a training program aimed at developing human abilities needed for the era of automation. One of the items stressed in the program was to promote interdependence and collaboration between school education and occupational training in industry. While its significance should be well appreciated, this program confined itself to a collaboration between upper secondary schools and industry. In my opinion, it is also necessary to institutionalize a cooperative system of training including the college level of education. In other words, in addition to providing upper secondary graduates with opportunities to receive a college education, it is desirable to make it a requirement for college students to gain practical training and experience in industry. This training should include not only those techniques necessary for the immediate work place, but ought also to provide students with basic experience of the several kinds of skills required if they are to become industrial leaders. With such a cooperative system established, colleges and universities, for the first time, would cease to be "ivory towers," and would become important contributors to the process by which students cultivate those abilities truly necessary for their future occupational careers. At the same time, students would learn to respect the experience of shop-level workers, as well as leadership skills in an industrial organization, rather than those abstract theories and empty ideologies they have so far been taught.

Unfortunately, these are still ideals that have not yet been realized in Japan. By and large, college education is very near to being a "mass production of diplomas," as it is often called, while secondary school education is very much like a preparatory school training.

Viewed in this way, it is doubtful that the younger generation in Japan today has a good prospect of making the leap, as far as training is concerned.

Are They Motivated for the Leap?

The third point to be considered is the extent to which young workers have a desire to make the leap.

It seems that the contemporary Japanese have a fairly strong drive for success in life, similar to that found in the "American Creed." In our survey of social stratification and social mobility quoted earlier, we asked a sample of Tokyo residents: "What do you think is the prerequisite for success?" The proportion of respondents giving "ability" and "endeavor" as the most important factors amounted to 54 per cent, which was more than the total of those who mentioned any of the other eight items given as the likely cause of success—luck, personality, worldly wisdom, educational career, family background, property, father's social position, and good connections with seniors.[11]

This testifies to the Japanese belief that those who have ability and work hard can, and should be able to, rise in the world, regardless of their family background or their fathers' social position. In contrast to the Japanese, Europeans

11. Odaka, ed., *Occupations and Social Stratification*, pp. 56–57.

still seem to value the idea of succeeding to their fathers' occupational status. Skilled workers in Europe, for example, appear to believe in the traditional principle of "like father, like son." This does not necessarily mean that they are resigned to low status, however; their pride in having been skilled workers for generations, together with their satisfaction at their relatively comfortable level of living, seem to be included in this value. In the case of Japanese workers, on the other hand, reality is more severe; they may not be able to survive if they follow the principle of "like father, like son," since keener competition has resulted from the heavy pressure of overpopulation. It is this that appears to underlie their strong aspiration for rising in the world.

At any rate, it is my impression that the Japanese desire for success is as strong as that of the Americans. They differ only in that the Japanese desire to rise often turns into a bitter resentment against persons who have succeeded and also in that ambition is usually relatively short-lived and more modest. These characteristics, and particularly the latter, can be found in the outlook on life of younger Japanese today. They are, however, better at calculating their chances and much more worldly-wise than their elders, who were brought up in the prewar years. As a result, they seldom cherish great ambitions, as some of the prewar generation did. Most have in their minds a life goal, not very large but surely accessible, such as securing a job in a first-rate company, creating a home with a pretty wife in a neat house, raising two or three children, obtaining such consumer durables as a washing machine, a color television, room coolers, and a car, and becoming a section head within fifteen or sixteen years.

They do not necessarily aspire to the position and role of top management. Though they are more or less critical of managerial policy, they think it wise to let management worry about the operation of the company. Once having secured a post in a large firm, an "escalator" called the *nenko* system will automatically carry them up at least to a rather high position. A secure life of petit-bourgeois comfort, with the guarantee of a slow but steady promotion, is the limit to which the calculating young generation aspires. Especially is this true of those lucky people who have obtained a white-collar job in a large industrial firm.

The workers employed in medium and small enterprises naturally have a different view of life. Skilled younger workers in such companies, for example, tend to disapprove the *nenko* system, particularly if they are talented and have high aspirations. What they desire, however, is not necessarily to become a top manager of a large firm. Their life goal, on the whole, is not very different from that of their brothers employed in large companies.

This outlook on life by the Japanese younger generation reminds one of that of the "organization man" described by William H. Whyte, Jr. The organization man is the type of man who is deprived of his autonomy and the opportunity for self-realization within a large, bureaucratized modern organization; he will nevertheless accept with resignation his self-estrangement, and try hard to maintain his position, or, though inwardly discontented with his circumstances and the managerial policy of the organization, will attempt to adjust himself to being a mere "cog" within a complex machine, and to seek in it a place of peace. According to Whyte, people of this type are rapidly increasing in present-day America, particularly among staff members and the middle level of management in all kinds of modern organizations.[12]

Assuming that this has been one of the major social trends in America since the 1940s, it is not surprising that a similar type has been appearing in increasing numbers in Japan since

12. William H. Whyte, Jr., *The Organization Man* (New York: Doubleday, 1956), chap. 1.

the 1950s, as the incidence of large, modernized organizations has grown, and with the advance of technological innovation. Since this is the case, what are we to do in order to cultivate future industrial leaders from among those well-rounded, peace-at-any-price "cogs" of the younger generation?

Before entering further into this discussion, however, let me refer to another common tendency found among younger people in Japan. I have pointed out that they are more or less like organization men, clever at calculation and seeking for a petit-bourgeois comfort in life. Another characteristic, overlapping with this in one sense, but quite in contrast to it in another, is that most younger people are "other-directed," to borrow the words of the American sociologist David Riesman. In *The Lonely Crowd,* Riesman asserts that in a society like that of present-day America, with its relative increase in those engaged in the tertiary industries, that is, sales, transportation, communication, and service, and where the natural growth of the whole population declines, there is a tendency for the "other-directed" type of people to proliferate. According to him, this type of person is best represented by the younger, "new-middle-class" generation working in metropolitan areas. In brief, the "other-directed" individual is a type of person who seeks his standards of judgment and conduct of daily life not in his own creed, but in the standards adopted by his contemporary "others." By contrast, a person whose standards are based on his own belief is called "inner-directed." The rugged individualists of former days, with a frontier spirit and the entrepreneurial mind, are a good example of the latter.[13]

Careful observers surely will have noticed that youngsters of the other-directed type are prevailing in postwar Japan as well. Such factors as the advance of mass consumption, the

---

13. David Riesman et al., *The Lonely Crowd: A Study of the Changing American Character,* abridged ed. (New York: Doubleday, 1955), chap. 1.

permeation of modern living in apartment complexes, and the development of mass communications have helped develop this tendency. In Japan today, the proportion engaged in the tertiary industries is already larger than that in either the primary or the secondary industries, though still small as compared to the situation in the United States or Great Britain.[14]

Moreover, Japan is a country where the social pressure of a group on its individual members has been traditionally very strong. It should be noted that even in prewar years there was a marked tendency for Japanese individuals to make decisions and to act based upon the patterns of value and conduct of other members of the group to which they themselves belonged, rather than upon their own belief and will. The other-directedness which can be a trait of the organization man has a good deal in common with this tradition. In consequence, the younger generation today is even more accustomed to act collectively and is still more apt at group discussions and circle activities, than the prewar generation of Japan.

At the same time, however, in the contemporary social and political environment there arise a number of conflicts between the groups and interests with which younger Japanese identify themselves. They are organization men under the control of their business firms, but they are also obedient to collective decisions of political parties and action groups which are often in conflict with the government and capitalistic businesses. The younger generation is thus often faced with a role conflict of extrinsic origins, due to the confrontation of groups, to both of which they owe loyalty.

On the other hand, they are also plagued by an internal conflict in their loyalty to one and the same organization. Submissive to their company, they also feel alienated in it.

14. In 1965, the percentages in the primary, secondary, and tertiary industries in Japan were 24.7, 32.4, and 42.9, respectively.

Obedient to a political action group, they sooner or later come to be alienated by its violence, the internecine struggles for power which characterize it, and the powerlessness of most members to influence its decisions.

The young are thus other-directed, but pulled and hauled by ambivalences and role conflicts, and often in consequence they engage in bizarre activities that seem completely to contradict any generalization one can make about them. For example, at the time of the public demonstration against the government in 1960 in connection with the Japan–U.S. Security Treaty, it seemed an everyday experience for young workers and students to spend day and night in zigzag parades, singing labor songs under red flags, and sitting down in front of the Diet Building throughout the night. For members of the older generation to take part in the same demonstrations, however, required an almost desperate act of resolve. That the young are used to collective action and tend to obey almost uncritically decisions previously reached collectively can also be seen in the more recent student revolts in their universities.

As another example, managers who have had bitter experience of workshop struggles organized by a labor union often complain that they cannot understand the younger generation. They ask how these college-educated employees, intelligent and well-behaved gentlemen, who even seem to understand the problems and responsibilities of management, can suddenly change into a rough and obstinate mob, wearing white towels around their heads, opposing management in every way, once they are ordered to take part in a workshop struggle.

Though I sympathize with these managers in their shock, their failure to understand this sudden change of attitude on the part of the young means that they have neglected to study the feelings and problems of the young. In my opinion, a polarity of attitude and behavior is inherent in young people today. It is of little utility for a manager simply to believe that he was betrayed by his younger workers. If on the contrary he makes earnest efforts to understand their problems and motivations, he might conceivably take actions that would lessen the difficulties which both he and they face.

In this connection, it may be well to have a look at the kinds of complaint and dissatisfaction younger workers have in large-scale enterprises in Japan.

In the case of medium and small companies, the main causes of worker dissatisfaction, according to *The Life of Young Workers* quoted earlier, are low wages and long hours of work, poor equipment, arbitrary management, favoritism by seniors, and an inferiority complex felt by young workers toward members of their age group who are lucky enough to be able to enter college or to obtain a job at a large company.

One of the major complaints of young workers in large companies, on the other hand, stems from the seemingly unfair distribution of wages, rather than their absolute amount. Dissatisfaction of this sort occurs particularly often in newly built automated plants, since there usually is a marked difference in work productivity between them and older plants of the same company. In the automated factory of the steel manufacturing company where we conducted research, for instance, the young workers newly employed there were able, mainly due to the automated arrangements, to produce almost five times as much as those who worked on a similar product at the older factories. Despite this difference in work efficiency, wages at both plants were set at the same standard, since the management thought it better to maintain the same wage system throughout the company.

In addition to the dissatisfaction with wages, a majority of young workers complain about the limitation placed upon them by the promotion system and the way in which they are transferred, which they think unfair or inappropriate.

Their expectations, and disillusionment, with regard to the "Blue-Sky" promotion system, cited earlier, is a good example.

The principal source of young workers' frustration in large-scale enterprises, however, is that, as mechanization and bureaucratization progress, workers are becoming more and more "small cogs in a vast machine" and so come to feel themselves alienated. In other words, such circumstances tend to cause the workers' dissatisfaction, because they have less chance to realize themselves, experience more difficulty in taking part in managerial decisions, and have to work like robots, deprived of autonomy and initiative. Even if they are organization men who are well-adapted themselves to a large, bureaucratized organization, they are still at least inwardly critical of management. Moreover, their circumstances, which allow them no way to vent their feelings within their work place, are likely to lead them to destructive counter-activities.

It is sometimes suggested that automation will bring about an improvement in the conditions which give rise to worker alienation. I myself am one of those who, in principle, count on this possibility. For one thing, since automation requires workers with higher skills and educational background, the social distance between workers and their superiors will be greatly reduced, in comparison with the gap in older-type factories, with the result that there will be more chance for workers to communicate with their superiors and to participate in managerial decisions.

Granting that this opportunity is possible, however, there is no guarantee that it will be realized with inevitability. Unless there is a conscious effort on the part of management to improve the situation, and to encourage workers' self-realization, the expected effects will never be forthcoming.

How, then, do workers vent their dissatisfaction? If there is no way for them to realize themselves or to recover their autonomy and initiative within the work place, where do they try to find it? In the case of young Japanese workers, the most common solutions for both blue- and white-collar employees are in such leisure activities as fishing, mountain climbing, skiing, or motorcycling; and such amusements and hobbies during their leisure hours as watching television, enjoying music, playing *go* and Japanese chess, mah-jong, pinball, or bowling. In prewar years, they were likely to be spoken ill of if they openly gave themselves over to fun even on holidays, and their seniors would look upon them as frivolous. Because of the postwar democratization, they now can act much more freely. Some engage in group activities for like-minded persons, or in publishing the magazine of an association; others take part in such activities of the local community as boy scouts, religious associations, or public welfare services. One positive solution may be to become active, or even to become a leader, in the local labor union, by which course they fulfill wishes and ambitions that are frustrated within the company.

At any rate, in this way young workers' dissatisfactions may at least partly be relieved. As a result, their complaints and criticisms may not have to rise to the surface. And for this very reason, managers tend to be little concerned about employees' discontent. It is probable that most industrialists in Japan do not realize how deeply dissatisfied young workers are. Only when the workers lay aside their usual meek appearance as organization men, and take sudden and violent collective action, does management become aware of the strength of their feelings.

To supplement the discussion above, let us look at some of the results of the employee attitude surveys I have conducted since 1952 at a number of large enterprises in Japan. The surveys, known in Japan as Workers' Allegiance Surveys, are, unlike ordinary morale surveys, designed to measure the degree to which workers are satisfied with both their company and the labor union at the same time. From the combination of their attitudes, we developed several categories and calculated the proportion of employees corresponding to each.[15] For convenience of illustration, let us call a category which is high or positive in the degree of both satisfaction with union and company "Pro," and another which is low or negative in the degree of both "Con." Thus, Pro type workers are those who have a favorable opinion of both company and union, while Con type employees are those who are dissatisfied with, or critical of, both.[16]

According to the surveys, throughout a wide variety of plants in the nine different companies studied, there were more of the Pro type among workers over forty, and more of the Con type among employees in their twenties. In Kokan Steel Tube, a large steel manufacturing firm, for example, of the employees over thirty-five, 38 per cent were Pro and only 6 per cent Con, while among younger workers the proportion of those who were Pro diminished to 22 per cent and that of those who were Con increased to 11, almost twice as many as in the older group. Similar results were obtained from my study of Shikoku Electric Power, a leading electric power company. Whereas 23 per cent of its members were Pro and 21 per cent Con among the older age group in their forties, only 12 per cent were Pro and as many as 35 per cent were Con among employees in their twenties.

When we compare the proportions of the two types by workers' educational career, there was a general tendency for those with more schooling to be more frequently of the Con type. Among upper secondary graduates in their twenties in particular, individuals in the Con category were remarkably numerous. In Shikoku Electric Power, for instance, Pro types

15. For the details of these attitude surveys, see Chapter Four, section on "Company Allegiance and Union Allegiance."

16. The original names of these types were, as will be seen in Chapter Four, "Pro-Pro type" and "Con-Con type," respectively.

decreased systematically in proportion to schooling: 26 per cent for graduates of primary schools, 19 per cent for those from lower secondary, 11 per cent for upper secondary graduates, and 10 per cent for those from colleges and universities. By contrast, Con types increased also systematically with education, the percentages being 16, 21, 32, and 37.[17]

What is important in this connection is that in those companies and plants where technological innovations have been introduced, there is a tendency for a higher proportion of the employees to be young, with higher educational attainment, and with attitudes of the Con type. According to the survey we conducted in 1960 at the Mizue Ironworks of Kokan Steel Tube, a newly built automated plant, for example, there were almost twice as many upper secondary or college graduates as at other plants of an older type in the same company, and there was a correspondingly larger proportion, 17 per cent, of Con type workers than at the other plants, where individuals of this type constituted less than 10 per cent on the average. The results from my 1961 survey at Tokyo Electric Power also show that persons of the Con category amounted to 16 to 28 per cent of the employees at each of the company's new automated power plants, as compared to only 12 per cent for the company as a whole.

What do these facts mean? As was stated above, a person of the Con attitude is critical of, or discontented with, both his company and union. He is likely to bear grudges against the company for its working conditions, welfare facilities, or personnel management; at the same time they are likely to be critical of the union's recent activities, administrative practices, or executives. The surveys show that people of this type are more frequently found among young employees, especially in their twenties, with an upper secondary edu-

17. The percentages shown here were obtained from our surveys at Kokan Steel Tube and at Shikoku Electric Power conducted in 1956 and 1957, respectively. See Chapter Four, Table 4.

cation, and that they are significantly numerous at those establishments where automation has been introduced.

It is to be noted that people of this type are not necessarily "destructive" or "rebellious," as they are often thought to be by management. More often than not they are merely critical or frustrated. One should bear in mind that highly educated workers tend to be critical of anything they are interested in, and that if they are young, their expectations with regard to the policies and activities of their company or union are apt to be very high.

If management fails to show insight into their psychology, and, on the assumption that Con type employees are "destructive" elements, takes the offensive to suppress or reject them, it will merely contribute to making them more critical or opposed. What is worse, such an offensive may turn at least a portion of such employees into really rebellious elements. One should not forget that there are those among Con type workers who, despite their intrinsically strong sense of belonging to both union and company, are still critical of the present state of both, chiefly because of their high expectations.

True, people of this type, relieving their discontent in their leisure hours, may not easily resolve upon a direct expression of their feelings at the level of behavior. Being used to collective action, however, they may suddenly turn to violent group resistance, whenever circumstances favor it. Once embarked upon such a course, moreover, most would feel that they ought to follow the decisions collectively made, even though they might not be wholly satisfied with them. People's reactions to a question posed in the surveys cited above may be taken to illustrate this point. The respondents were asked whether a man should follow the decisions of his union even if he were not convinced that they were proper. In the case of Shikoku Electric Power, only 22 per cent answered, "One need not if one is not convinced," while 54 per cent replied, "One should even if one is not convinced." Particularly noteworthy are the proportions of the latter answer by age groups. Whereas 56 per cent of the employees under thirty answered, "One should," only 37 per cent of those over fifty gave the same answer, those replying, "One need not," increasing to as many as 36 per cent. This will testify that younger workers generally are more ready to take collective action than are their elders.

How Ready Is Management?

In the foregoing I have discussed whether younger people are now prepared, objectively as well as subjectively, to make what Drucker calls the "leap." How ready, then, are the present industrial leaders to help them to take this step?

It is customary for top Japanese management to declare that success in business depends on the human element. Particularly in a country like Japan where natural resources are scarce, they say, human resources are all the more precious. The effective use of human factors, as well as the development of human abilities, are the prerequisites not only for the future prosperity of an enterprise but also for the successful economic growth of the country as a whole. These, they declare, constitute the industrialist's highest mission.

The question here is how well this mission has been carried out. Fortunately or unfortunately, human resources are in excess in Japan. As a result, there still is a tendency for industrialists to value machines and equipment above human factors. Despite the recent fad of "human development," it remains in most cases merely empty talk.

There are of course many things that industrialists alone cannot possibly reform, even if their intentions are good. Among the younger generation, for example, there are many poor individuals who, despite real natural ability, have no chance to attain a position of industrial leadership. Although it is vital for such individuals to have the opportunity to enter a good school and to climb up the social ladder without serious hindrance and discrimination, the reforms needed here are probably beyond the power of private management. To cope with the situation will require the government's support, the enlargement of scholarship funds and, above all, a reform of the educational system and of school facilities.

Among industrialists there also are those who cannot afford to help the younger generation, mainly because their own firms are in adverse circumstances. It is unreasonable, for instance, to expect much in this line of the poor owner of a petty firm, or of the management of a declining company facing a large curtailment of personnel.

Excluding such cases for the time being, however, I want to confine myself to those actions by present industrial leaders which are prerequisites to the younger generation's advance, but which to date have been lacking or largely neglected. The following are the more important items I have in mind:

1. Lacking among present leaders are those who themselves set a good example for future leadership, particularly as regards the abilities and enthusiasm they devote to managerial reform. No doubt some industrialists are indeed far ahead of intellectuals in the originality of their ideas, their ability to plan, and their powers of execution. They are in a small minority, however, and the greater number of managers seem to be drones who, without the qualifications or will to be real leaders, strive merely to keep up their present position and power. Very often the latter are indifferent to the choice of able young men.

2. Such superficial factors as formal educational career or personal connections, rather than the individual's real abilities and talents, are still considered the primary criteria in hiring employees. It is understandable that management

should consider it less of a risk for the company to hire those who graduated from a top-ranking university, or who have special connections with company executives. In so doing, however, it overlooks a large proportion of really able persons, and in their place adds a certain percentage of future deadwood. In this connection, industrialists should consider the baneful effects of the system of mechanical selection and promotion, according to which college graduates, whether talented or not, are always given priority over secondary school graduates, not only for employment and initial wages, but also for their future promotion. For one thing, this is the main cause of the tendency by parents and teachers to make too much of a child's formal educational career. This pernicious practice will be reformed if management will only change its policy and regard a young man's promise and real abilities as the first requisite for selection.

3. Generally neglected, or performed in only a perfunctory manner, is training for future managers and skilled workers of a higher level. Present industrial leaders seem to feel it "inefficient" to spend time in employee training, even when it is actually needed. Very often they overlook the fact that lack of training causes workers to suffer from low morale, unnecessary tension, and mental fatigue, which in their turn result in a larger loss of productivity. Even when they do devote time to the education of employees, there is a general tendency for industrialists to treat it in terms of school-like practices, with trainees being expected to take the passive role of students attending a lecture. In actuality, this type of education can form only a part of employee training. In order to cultivate competent staff officials of the future, it will be necessary to let candidates undergo a more practical type of training using such techniques as the "case method," "role playing," or "sensitivity training." Even more important for bringing up future management, however, is to make candidates learn through actual experience, by giving

them ample opportunities to take part in managerial decisions and responsibilities.

4. Still more neglected on the part of the present leaders is the practice of treating employees as their partners. I do not necessarily mean that workers are not regarded as human beings under the Japanese managerial philosophy. Unlike their counterparts in Western countries who are used to treating workers according to an impersonal logic of efficiency, Japanese businessmen are basically paternalistic and never consider employees merely as a labor force or as tools. In this respect, it may be said that Japanese businessmen have been well acquainted with the "human relations" approach for a long time. On the other hand, however, the practice of dealing with employees as their partners has not been known to them, at least until recently. It has been customary for them to regard, and actually to treat, workers as human beings belonging to a lower social grade, to be considered their "servants," or helpless children in constant need of parental care and supervision. So long as they preserve such a traditional view, however, it is unlikely that they will be able to cultivate capable leaders in the future. The present business leaders should at least realize that many of their employees nowadays have a higher educational attainment and are better equipped with technical knowledge than themselves.

5. Also lacking is an adequate understanding and consideration of the feelings of the younger generation on the part of the present leaders. In spite of their paternalistic disposition, Japanese industrialists appear to be quite indifferent about the feelings of their employees, and especially of young workers. Their indifference, however, takes at least three forms, namely: a failure to notice how the workers actually feel; making light of their employees' sentiments; and overconfidence of their ability to understand, even when they do not. Of these, the most harmful is obviously the third. There seem to be a great many industrialists who, largely because of

this kind of indifference, unconsciously hurt the feelings of young workers and thus foster their hostility. They do not, for example, understand the psychology of young people who, though as individuals they may act like yes-men, can easily be transformed into a rebellious mass in collective action.

6. Predominant are types of personnel management which are now suppressive, now laissez-faire, and now manipulative. Especially is this true of the techniques aiming at heightening the work morale of young workers. A truly democratic approach, which would channel young workers' energy away from resistance and into managerial reform, is still sadly lacking. Take, for example, the case of younger workers of the Con type who are discontented with both company and union. To deal with their resistance, there are roughly four different approaches, as follows: suppressing them within the framework of the existing managerial system; leaving them to find a natural solution with the passage of time; converting them into submissive, superficially company-oriented employees by the use of manipulative techniques; and leading them to an active participation in an organizational reform by directing their energy for resistance into that channel. Of these, I can recommend only the last as an approach that will be effective in terms of both worker satisfaction and the development and prosperity of the organization. Executives today should be aware that young workers resist because they feel their self-realization will never be satisfied under the existing managerial system. In other words, since they are frustrated and feel themselves alienated, they resist management. If they are given the chance to realize themselves and to enjoy autonomy in their work place, this antagonism will dissolve into a more constructive attitude.

7. Still underdeveloped, finally, is the practice of selecting employees who have the makings of future leaders, letting

them undergo special training and trials and boldly promoting them to posts of special importance. It goes without saying that even a system of unlimited promotion cannot make all employees department heads or division chiefs. It is also obvious that however excellent the training it cannot make future leaders out of persons who lack natural talent and ambition. It is in fact surprising that, while competition is bitter and rampant outside the enterprise in Japan, within it there is very little competition among employees. Except for the ranking by educational career at the time of employment, everybody in a firm, whether talented or not, diligent or idle, is treated almost alike, promoted primarily by seniority, and entitled to remain peacefully in the same firm until retirement. This harmonious atmosphere, which is truly common in Japanese establishments and in which competition among individuals is discouraged, has of course a long history. In brief, one may say that the traditional managerial philosophy, which gives precedence to the well-balanced development of the organization as a whole over the competitive self-realization of its members, as well as the *nenko* system based on that philosophy, are responsible for creating, or at least preserving, this atmosphere. In the postwar years, the labor union movement, in the name of fair livelihood opportunity for all workers, has opposed the introduction of any new system which would treat employees on the basis of ability and achievement. At any rate, and under such circumstances, many executives cannot easily bring themselves to pick out talented employees and promote them to posts of importance, or, on the other hand, to shift the deadwood into unimportant posts, or fire them. If it is of the utmost importance to encourage well-qualified young workers to leap the generations, what the present leaders must do first of all is to initiate a thorough reform of the long-established system of employee promotion and the underlying philosophy of traditional management.

Viewed in this way, it is clear that preparation for "the leap of the younger generation" is a responsibility primarily of present business leaders, rather than of the younger generation itself. Never before has the improvement of leaders' quality and the establishment of a really effective leadership been as badly needed as it is today. Perhaps the present leaders, even more than those of the future, urgently need intensive training designed to help them meet the requirements of the era of automation.

CHAPTER FOUR

Workers on Their Allegiance
to Union and Company

Five Aspects of Industrial Relations

There are at least five different aspects to industrial relations: (1) class relations, for example, those of the working class versus the capitalist class; (2) organizational relations between labor unions and management; (3) employment relations between employer and employees; (4) job relations, such as those between shop supervisors and rank-and-file workers; and (5) role relations, that is, the relations between the two different roles of employee and union member held by an individual worker.

The first aspect, which may be variously described also as the proletariat versus the bourgeoisie or the working masses versus the industrial elite, manifests itself in both the local community and the national society as a whole. It is evident that such class relations are, objectively and psychologically, related to status differences between workers and management within the enterprise.

The second aspect of industrial relations equals what is usually called "labor relations" or "union–management relations." This most common usage of the term designates relations between a company as an organization and a union as its counter-organization. The union as an institution aims intrinsically at protecting workers' rights, and seeks to improve living standards and the social status of its members. Its objectives and functions not only differ from, but also are in some important respects entirely opposed to, those of a company. In other words, it is the company's "social opponent." Industrial relations in this second aspect may be characterized above all as relations of opposition.

The third aspect of industrial relations, which is commonly called "employee relations" or "employer–employee relations," is, by its very nature, an intra-organizational phenomenon. It consists of formal and informal interpersonal ties and corresponds to what is often termed "human relations in industry" in a broad sense. The formal rules and practices instituted under the name of "personnel administration" or "labor management" regulate this aspect of industrial relations. However, no matter how extensively the rules and practices are formalized, informal interactions within an organization between employees and managers as individuals will still remain. Again, this aspect may be characterized primarily as cooperative relations, as against those of opposition previously outlined. The cooperation may not necessarily be spontaneous or voluntary. It is incontestable, however, that a worker's role as an employee is that of a cooperator, while he may, at one and the same time, be a member of a labor union which is essentially a social opponent of management.

The field within which the fourth aspect, namely, job relations, manifests itself is still narrower than that of the third; it emerges at the shop level of a company or a plant. Although it consists primarily of the relations between shop supervisors and rank-and-file workers, this aspect in practice is much more complicated than is commonly assumed, despite the narrow focus at which it emerges. Job relations are not only inseparable from employment relations but are also exposed to the direct pressure of union influence. Moreover, one should conceive of them as webs of relationships involving not only shop supervisors and operators but also other on-the-spot and behind-the-scene people, such as company executives, middle management, union leaders, and shop stewards.

Until very recently, students of industrial relations have largely dealt only with the foregoing four aspects of the problem. In fact, these may well be taken as sufficiently exhaustive, if we view industrial relations as being essentially relations between two separate objective entities. However, if we attempt to analyze industrial relations with more penetrating insight into the psychology and behavior of individual workers, we need to locate our analytical approach closer to the workers' standpoint. To do so, in addition to those already noted, a fifth, namely, the relationships between the two different roles of an individual worker, should also be an object of attention. Industrial role relations, in this sense, refer to the relationships between a worker's role as an employee and the role expected of him by virtue of his union membership.

These two roles, though they are performed by a single individual, are not necessarily compatible with one another. Instead, at least theoretically, the two are mutually exclusive, for they indicate a worker's obligation to belong to the two different organizations—company and union—of which the relationships are essentially those of opposition. A worker may be able to satisfy his needs by virtue of his membership in both. Yet, he is exposed to the influence and restrictions of these two organizations, which stand against each other in terms of their organizational objectives and interests. Individual workers are, so to speak, under the influence of two diametrically opposed gravitations. The power of the two will increase when the opposition between the two organizations becomes open. It may even become unbearable for individual workers when the opposition grows to the acute stage of "struggle relations," as in long, hard strikes or bitter workshop struggles. Under such circumstances, when the contradiction between the two roles becomes intense, a worker must face a difficult choice. He may be compelled to betray his fellow unionists and to act against the union leadership in order to maintain his "employee status." Or, he may have to violate company rules and openly oppose his superiors, even at the risk of being discharged, in order to be loyal as a "fighting unionist." To be sure, such a circumstance is the extreme. Even in a more normal situation, however, one cannot deny that the two gravitations continue to influence, directly or indirectly, the psychology and behavior of individual workers.

It should be noted, in this connection, that the degree of concern for, and interest in, two roles of many workers constitutes one of the basic factors at least indirectly determining the nature of union–management relations. If, for example, we find that a majority of workers concentrate their attention on their role as union members, while neglecting that as employees, or if the opposite is true, we can reasonably expect that union–management relations in these cases will be unstable. If, on the other hand, we find that a majority of workers show positive concern for both roles, we can conjecture that the nature of union–management relations on the spot will be very different. Needless to say, the nature of union–management relations at a given time is not determined solely by the intentions and feelings of the working people involved. There are other important factors affecting these relations, such as historical and environmental conditions, general views and attitudes of the leaders on each side, the nature of their strategies, the quality of their leadership, and so forth. Nevertheless, we must admit that workers' concern for the two roles—which will be referred to hereafter as "union allegiance" and "company allegiance" —constitutes one of the basic factors determining the nature of union–management relations.

The five aspects of industrial relations briefly described above are in reality closely interrelated. Accordingly, no matter which one we select as the focus of analysis, the requirements of scientific analysis of empirical settings make it necessary for us to take the other aspects into account as well. Admittedly, there are several analytical approaches to industrial relations. For example, labor economics as well as labor law studies tend customarily to view the subject from the second point of view, namely, as union–management relations. The studies of business administration concern themselves primarily with the third aspect. Almost no attempts, however, have so far been made by these disciplines, which have been dominant in studies of industrial relations, to approach the subject in terms of the fifth aspect.

Under such circumstances, I propose to take workers' allegiance to union and company as the problem for the present chapter. In the following pages, I shall attempt, based on the data obtained from our surveys, to analyze industrial relations through a closer examination of the worker's concern for his two roles.

Company Allegiance and Union Allegiance

The term "allegiance" is used here for a "sense of belonging" to a group. It may be defined as the degree to which a member of a group or organization is psychologically inclined to accept it as "his own" and, therefore, to regard himself as being "one of the group." Belonging to a group, in this context, means more than formal membership. The statement that an individual is highly loyal to a group means not only that he is a member in objective terms, but also that he sees the group subjectively as the basis of his own daily activities and that, in turn, he identifies himself primarily with that particular group.

Secondly, the term presupposes an individual's preference for the group to which he belongs. In the case of organized workers, therefore, the question of allegiance becomes significant to industrial relations in the following terms: which of the two organizations, union or company, do workers tend to prefer as their own basis of daily activities? Or, to put it differently, to which of the two organizations are workers psychologically inclined to belong? In short, are they "pro-management" or "pro-union"? On the basis of their preference, we may construct "allegiance types," which will be discussed below.

Finally, the term "allegiance" not only implies discrimination between union and company, but also may be quantified by degree. Needless to say, the degree of allegiance varies among individuals. If it is found to be higher than average, I call it a "Pro" attitude. If, on the other hand, it is lower than average, it should be understood to signify an antagonistic or critical attitude toward management or union, and I call it a "Con" allegiance. Between the two, one can conceive of a middle range of impartial or indifferent attitudes toward company or union, which I call "Neutral." In this way, it is possible to classify all the workers under survey into Pro, Neutral, and Con types. Company allegiance will be referred to here as Pro-M, Neutral-M, and Con-M, and union allegiance as Pro-U, Neutral-U, and Con-U, where "M" stands for management to represent company, and "U" for union. The six types resulting from the foregoing classification may be called "simple allegiance types" to distinguish them from "combined allegiance types," which will be explained in the following.

Company allegiance and union allegiance, which characterize the worker's concern for his two different roles, are generally considered to be in conflict with one another. Thus, if company allegiance of a worker or group of workers is found to be high, it is customarily conjectured that his or their union allegiance should be low, and vice versa. In fact, many intellectuals unacquainted with the realities of industrial relations adopt such a stereotype. Moreover, the leaders of both unions and companies also seem to hold a similar view. As will be seen later, however, the way workers actually think can differ considerably from that assumed by such a stereotyped and superficial view.

To clarify this point, we must examine how the worker's concern for union and company are combined with each other, since the two in actuality are closely intertwined within the consciousness of a worker. Classification of workers in terms of the "combined allegiance types," which will be obtained by combining their company and union allegiance, was designed to serve this end. Theoretically, the combinations of company and union allegiance, each with the three types Pro, Neutral, and Con, result in nine possible combined types. Of the nine, five show distinct characteristics and should, therefore, be accorded special attention. These five include Pro-M-and-Pro-U, Pro-M-but-Con-U, Neutral-M-and-Neutral-U, Con-M-but-Pro-U, and Con-M-and-Con-U, which, for the sake of convenience, will be abbreviated to Pro-Pro, Pro-Con, Neutral-Neutral, Con-Pro, and Con-Con, respectively.

The Pro-Pro type, indicating high allegiance to both company and union, can be called "dual allegiance." The Pro-Con type, denoting those workers who identify themselves with the company but do not support the union, may be called, for lack of a better term, "unilateral company allegiance." The Neutral-Neutral type, tending to be in the middle in their attitudes toward both company and union, can be called "nonpartisan." The Con-Pro type, designating those who do not support the company but identify themselves with the union, may be called "unilateral union allegiance." Finally, the Con-Con type, indicating those who are antagonistic to, or critical of, both company and union, may be tentatively called, again for lack of a better term, "discontented." The remaining four combinations out of the nine—Pro-Neutral, Neutral-Pro, Neutral-Con, Con-Neutral—are too ambiguous in their characteristics to be treated as distinct types. Moreover, cross-tabulation with other attitudinal items revealed a similarity of inclination among them. For these reasons, they will be treated collectively as "others." Table 1 shows the names and brief descriptions of these types.

Table 1

Allegiance Types

Simple Allegiance Types

Pro-M	Positively identify with company	Pro-U	Positively identify with union
Neutral-M	Assume a neutral attitude toward company	Neutral-U	Assume a neutral attitude toward union
Con-M	Dissatisfied with company	Con-U	Dissatisfied with union

Combined Allegiance Types

Pro-Pro, or Dual Allegiance	Have high allegiance to both company and union
Pro-Con, or Unilateral Company Allegiance	Pro-company but do not support union
Neutral-Neutral, or Nonpartisan	Have a neutral attitude toward both company and union
Con-Pro, or Unilateral Union Allegiance	Pro-union but do not support company
Con-Con, or Discontented	Dissatisfied with both company and union
Others	Fall into the types Pro-Neutral, Neutral-Pro, Neutral-Con, and Con-Neutral

A series of surveys to measure allegiance types as described above have been conducted, under my guidance, by the Sociology Department of the University of Tokyo since 1952. Similar multiple-choice questionnaires designed to elicit workers' attitudes toward union and management were administered to some thirty thousand employees working at more than seventy plants belonging to nine large-scale Japanese firms. The nine include Kokan Steel Tube, Kogaku Optics, Shikoku Electric Power, Tokyo Electric Power, Matsuya Department Store, and Okamura Manufacturing.[1] The surveys usually defined the universe as being all employees below the *kacho* (section head) level. When sample studies were made, a stratified random method was employed, with sample size ranging from 1/19 to one half of the universe.

The questionnaire items usually included questions about: employees' background, such as age, school career, status in the firm, and so on; opinions of the job, facilities, human relations in the workshop, and so on; reactions to wages, promotion system, welfare programs, and so forth; a choice between two opposing views on such topics as union–management relations, productivity increase, technological innovation, and so forth; and political inclinations, attitudes toward life in general, and the like. In addition, each questionnaire invariably included a set of twelve items, designed specifically to measure allegiance types. This set of questions was used in every case with no significant change of wording. Half of the twelve questions are concerned with the company—its working conditions, managerial policies, the quality of managerial leadership, its business performance, and so on. The other half are related to the union—the degree of its importance in improving working conditions, union policies, the quality of its executives' leadership, union

1. An outline of the surveys completed by 1967 is shown in Table 6 (p. 124).

activities, and so forth. The questions were prepared so as to classify employees' attitudes toward each of the two parties into five grades, on a scale ranging from very favorable to very unfavorable. Besides the uniformity of the twelve questions, much attention was paid to standardizing the method of data collection and the procedure for identifying the allegiance type of each employee, so as to permit ready comparison of the tabulated results drawn from the various plants and companies under survey.[2]

Studies similar to mine concerning worker allegiance have been carried out in the United States several times since 1950. At least five such studies have been published to date.[3] At the time we began the first survey at Kokan Steel Tube in 1952, however, these American studies had either not yet started, or not yet been reported. Since we embarked on our surveys independently, it should be added here that our studies have little in common with the American counterparts in regard to methods of data collection, design of scales for attitude measurement, classification of union and company allegiance, as well as the theoretical framework used for research.

2. The actual procedure for classifying employees according to allegiance types may be summarized as follows: First, the individual's response to each of the six related questions was classified into five grades, and then was assigned a prearranged score. In this manner, the total score for the entire six questions related to the company or the union was obtained for each respondent, ranging from 6 to 30 points, with the midpoint of 18. Next, this range of 6 to 30 points was divided into seven equally distanced categories of 3Pro, 2Pro, 1Pro, N, 1Con, 2Con, and 3Con, the first three of which were in most cases combined to represent Pro and the last three Con.

3. Arnold M. Rose, *Union Solidarity: The Internal Cohesion of a Labor Union* (Minneapolis: University of Minnesota Press, 1952); Theodore V. Purcell, *The Worker Speaks His Mind on Company and Union* (Cambridge, Mass.: Harvard University Press, 1953); Ross Stagner, *The Psychology of Industrial Conflict* (New York: John Wiley, 1956); Ross Stagner, Theodore V. Purcell, Willard A. Kerr, Hjalmar Rosen, and Walter Gruen, "Dual Allegiance to Union and Management: A Symposium," *Personnel Psychology* 7 (March 1954), pp. 41–80.

The Distribution of Allegiance Types

In the sections that follow, the research findings relevant to the present subject will be presented under three headings: comparison of the distribution of allegiance types among different companies and plants; structural characteristics of allegiance types; and worker attributes in relation to their allegiance types. In presenting these findings, I rely mainly upon the data drawn from the surveys at Kokan Steel Tube (hereafter referred to as Kokan Steel), one of the largest steel manufacturing firms in Japan, and at Shikoku Electric Power (hereafter Shikoku Electric), one of the nation's nine leading electric power companies.[4] The surveys were conducted at the former in 1952, 1956, 1960, and 1963, and at the latter in 1957, 1961, and 1965.

To begin with, let us take a look at the distribution of workers by allegiance type. Column A of Table 2 gives a general picture of worker attitudes toward company or union. Figures like 54, 47, 18, and 26 in the table represent the percentages of occurrence in the sample size, excluding the "don't know" and blank responses. The signs "M" and "U" in Column A headed "Inclination" show the relative strength of the simple types of allegiance to company and union, respectively—that is, Pro-M versus Pro-U, and Con-M versus Con-U. Column B on the right side of the table shows the distribution of combined allegiance types, such as Pro-Pro, Pro-Con, and so on. Figures under the respective combined types, such as 34, 7, 9, 3, and 10, are again percentages of the sample. "Others" includes, as was explained above, the four ambiguous types, Pro-Neutral, Neutral-Pro, Neutral-

4. The distribution of allegiance types in the companies and plants other than Kokan Steel and Shikoku Electric is shown in Table 7 (p. 127).

Table 2

Distribution of Allegiance Types in Two Companies
(Per Cent)[a]

Company and Plant	A. Company and Union Allegiance					B. Combined Allegiance Types							Number of Respondents	Distribution Pattern
	Pro-M	Pro-U	Con-M	Con-U	Incli-nation	Pro-Pro	Pro-Con	Neut.-Neut.	Con-Pro	Con-Con	Others	Total		
Kokan Steel (1952) Kawasaki Ironworks	54	47	18	26	M	34	7	9	3	10	37	100	(701)	X
Kokan Steel (1956) All Nine Plants	50	53	21	21	–	32	6	8	7	9	38	100	(1,861)[b]	X
Kawasaki Ironworks	56	55	16	20	–	37	8	9	4	6	36	100	(865)[c]	X
Tsurumi Ironworks	42	70	23	10	U	34	2	7	11	5	40	100	(630)	X
Toyama Ironworks	52	79	22	5	U	46	1	5	15	2	31	100	(363)	X
Niigata Ironworks	57	46	23	24	–	34	9	5	4	9	39	100	(92)	X
Koyasu Ironworks	32	64	37	10	U	30	0	10	15	9	36	100	(87)	X
Rozai Brickfield	25	85	36	2	U	24	0	7	29	2	38	100	(97)	Y
Tsurumi Dockyard	42	27	30	46	M	18	13	9	3	21	37	100	(267)	Z
Asano Dockyard	43	26	27	49	M	18	11	6	1	21	42	100	(187)	Z
Shimizu Dockyard	19	31	46	37	U	8	4	12	11	24	41	100	(92)	Z
Shikoku Electric (1957) Whole Company	30	27	42	45	–	15	7	10	5	27	36	100	(1,804)	Z

Company and Plant	A. Company and Union Allegiance					B. Combined Allegiance Types							Number of Respondents	Distribution Pattern
	Pro-M	Pro-U	Con-M	Con-U	Incli-nation	Pro-Pro	Pro-Con	Neut-Neut.	Con-Pro	Con-Con	Others	Total		
Head Office	17	16	61	53	–	9	4	10	2	43	32	100	(128)	Z
Tokushima Branch	39	27	35	48	M	18	11	7	2	24	38	100	(341)	Z
Kochi Branch	26	27	42	44	–	14	7	14	6	26	33	100	(388)	Z
Ehime Branch	33	32	35	38	–	17	6	10	6	21	40	100	(536)	Z
Kagawa Branch	28	23	50	50	–	14	6	10	5	34	31	100	(264)	Z
Steam-Power Plants	20	26	50	51	–	12	6	12	9	34	26	100	(147)	Z
Kokan Steel (1960) Mizue Ironworks	43	31	28	38	M	19	10	10	6	17	39	100	(1,051)	X'
Shikoku Electric (1961) Whole Company	29	33	41	35	U	17	3	10	7	23	40	100	(2,664)	Z
Kokan Steel (1963) Whole Company	26	26	44	47	–	13	5	11	5	29	37	100	(3,917)	Z
Shikoku Electric (1965) Whole Company	36	35	34	33	–	21	5	13	5	19	37	100	(2,638)	X

a Percentages show the frequency of occurrence to the sample size, excluding the "don't know" and blank responses.

b The number of respondents of all nine plants in Kokan Steel (1956) equals 2,000, randomly drawn from the total sample of 3,555, minus the "don't know" and blank responses.

c The number of respondents of each plant shown here includes that of shop workers only, whereas the number of respondents of all nine plants includes that of both office and shop workers.

Con, and Con-Neutral. The signs "X," "Y," and "Z" in Column B headed "Distribution Pattern" indicate differences in the pattern of distribution of the combined allegiance types, as is explained below.

From Table 2 we observe some common tendencies in the distribution of allegiance types among different companies and plants. Since all the figures under the section "Others" in Column B approximate 37 per cent, the proportions of workers falling into the five significant allegiance types, such as Pro-Pro and Pro-Con, if added together, should come close to the remaining 63 per cent. If this 63 per cent were randomly distributed, each of the five allegiance types would amount to about 13 per cent. Yet the fact is that the proportion of the Pro-Pro, or dual allegiance, type exceeds the 13 per cent level in a majority, that is, in 16 plants and companies out of the total of 20 shown in Table 2, and in as many as 21 cases out of the total of 27 shown in Tables 2 and 7. This means that the workers with dual allegiance tend to represent a greater proportion of the total worker population than do those falling in any other allegiance type, despite apparent differences in distribution by company and plant. Moreover, in 12 cases shown in the two tables, the dual allegiance type reaches more than 20 per cent. Where this is true, this pattern is referred to here as the "X pattern."

It does not follow, however, that the proportion of the Pro-Pro type is always greater than that of any other type. As is apparent from the tables, next to the Pro-Pro type comes the Con-Con, or discontented, type, scoring over 13 per cent in 17 plants and companies out of the total of 27. In 14 of these cases, the discontented type exceeds 20 per cent, and this distribution is referred to as the "Z pattern."

On the other hand, the other three types—Pro-Con, Neutral-Neutral, and Con-Pro—show percentages under 13 per cent in the majority of plants, exceeding it in only a few exceptional cases. In fact, more often than not, they fall below

10 per cent. Also, there are only 4 cases where either the Con-Pro or Neutral-Neutral type exceeds either the Pro-Pro or the Con-Con. Such a distribution pattern is referred to as the "Y pattern." The X pattern of distribution where the Pro-Pro type is dominant but with less than 20 per cent is called "X-prime pattern."

From the foregoing we may conclude that there is a marked similarity among the distributions of allegiance types for each company and plant in that either of the two positive correlation types—Pro-Pro or Con-Con—is in most cases predominant. Differences in the allegiance type distribution among plants and companies are no less conspicuous, however. For example, a comparison in 1956 of five ironworks with three dockyards, all belonging to the same company, Kokan Steel, reveals that in the ironwork group the X pattern of distribution tends to be common, while in the dockyard group the Z pattern is the rule. In other words, the Pro-Pro, or dual allegiance, type is numerous among ironworkers, whereas the Con-Con, or discontented, type is conspicuous among dockyard workers. Such a distinctive difference in the pattern of distribution is also found between the two companies, Kokan Steel in 1956 and Shikoku Electric in 1957, when each is compared with the other as a whole. Similarly, significant contrasts and differences will result if comparisons are made among other companies and plants.

Why, then, do such distinctive differences arise? This is not the place to answer this question in detail for each case. Here I shall confine myself to two examples. In Column A of Table 2, for Shikoku Electric in 1957 we find the greater percentages of Con-M (42 per cent) and of Con-U (45 per cent) as against Pro-M and Pro-U, resulting in a large proportion of Con-Con type workers (27 per cent) in Column B. In this case, the greater size of Con-M is likely to be due partly to the workers' dissatisfaction with working conditions, particularly with their wages. The company's wage level was conspicuously low among the nine major electric power companies in Japan, and this was well-known to most of the employees. Moreover, managerial policies tended to be impersonal and bureaucratic, with the result that employees did not have much chance to exert their abilities and talents, nor were they given any opportunity for self-government in the workshop.

The union of Shikoku Electric, on the other hand, had once been one of the major constituents of the militant Densan (the former All-Japan Electric Workers' Union). Subsequent to the dissolution of Densan as a result of its defeat in the long, desperate strike of 1952, however, the employees formed an enterprise-wide federation called Shikoku Electric Workers' Union. At the time of the survey in 1957, the union was affiliated with the Zenro (the former Japanese Trade Union Congress) which, in its turn, was later merged into the Domei Kaigi (Japanese Confederation of Labor). In the course of these events, the policies of the union leadership changed and became more realistic and, in a sense, more "moderate" than in the past. It appears, however, that not a small portion of the union members were dissatisfied with this "moderate" strategy and hoped for a return to the militant and action-centered approach which they had experienced during the Densan period. In the eyes of these discontented people, the low wage level of the company, as well as the "autocratic" way of its management, were ascribable to the lack of militancy on the part of the union. The greater size of the Con-U type may be accounted for in the light of this historical background.

By contrast, in the case of Kokan Steel in 1956, the union belonged to the Tekko Roren (Japanese Federation of Iron and Steel Workers' Unions) which, in its turn, was affiliated to the Sohyo (General Council of Trade Unions of Japan) and had proved its militancy and strength many times through frequent disputes. The company, on the other hand, proud of its long history as one of Japan's leading steel producers, ranked high in the top bracket in terms of working conditions as well. This was at least one of the reasons why in Column A Pro-M (50 per cent) and Pro-U (53 per cent) are much larger than Con-M and Con-U, and under Column B the Pro-Pro type is as high as 32 per cent.

The marked difference in the distribution pattern between the two companies may also be explained in terms of the fact that young workers with an upper secondary school background are a great proportion of the total employees in the case of Shikoku Electric. As will be seen later, it is a general tendency for those who are young and have relatively high educational qualifications to be the discontented type, rather than that of dual allegiance. At Kokan Steel, however, where the Pro-Pro type was predominant, the proportion of employees with an upper secondary school or higher education who worked at its ironworks at the time of the survey was only 28 per cent, whereas in Shikoku Electric, where the Con-Con type was numerous, this proportion was as high as 58 per cent. It seems clear, therefore, that the factors of age and educational career greatly influenced the survey results at these two companies.

Such distinctive differences between the two companies, however, seem to be declining in more recent years. As will be seen in Table 2, when the data for Shikoku Electric in 1961 and Kokan Steel in 1963 are compared, both companies show a similar Z pattern of distribution. In 1965 Shikoku Electric even exhibited an X pattern. These changes in the distribution pattern may be explained as follows. The general decrease in the Pro-Pro type and the corresponding increase in the Con-Con type in Kokan Steel during the 1960s seem to reflect a general tendency for workers to become more egocentric and therefore less concerned for both company and union. Moreover, the management's policy taken subsequent to the technological innovation started in the late 1950s, failed sufficiently to meet the demand of the workers. The union, on the other hand, became more bureaucratic in its leadership and an enlarged social distance developed between leaders and rank-and-file members after the formation of a company-wide federation during the same period. The gradual increase in Con-Con type workers in Kokan Steel may at least be partially explained by these changes.

In the meantime, the union of Shikoku Electric, with its realistic policy, was gradually gaining the support of the members. Especially since a general improvement in wages carried out under the pressure of the union during the early 1960s the union has established confidence among rank-and-file workers. At the same time, because of better treatment, the number of people who used to be dissatisfied with management gradually decreased. Furthermore, a new managerial policy emphasizing human development and employee participation, which was started immediately after the first survey conducted in 1957, have contributed to the increase of workers who identify themselves with the company. The change in the allegiance type distribution in Shikoku Electric seems to have resulted from these new developments.

Structural Characteristics of Allegiance Types

A worker's consciousness is not composed merely of his attitudes toward union and company. His inclinations and value orientations to a variety of other subjects are closely related to his company and union allegiance. The list of such related attitudes will include, for example, the degree of satisfaction with his job, the sense of belonging to his workshop group, his attitude toward labor–management cooperation, his political party support, and his general view of life and the world. By discriminating among the patterns of these related attitudes for each allegiance type, we should be able to attain a more realistic insight into the structure of worker allegiance.

Table 3 shows, based upon the data drawn from my surveys at Kokan Steel Tube in 1956 and Shikoku Electric in 1957, the results of such an attempt at structural analysis. In this table, the respondent's related attitudes are classified into eight categories, as follows: (1) job satisfaction, which includes the degree of the respondent's liking of his job, and his evaluation of its importance; (2) satisfaction with the workshop, including the actual cooperation perceived among the members of his workshop group, as well as the perceived fairness of his direct superior; (3) philosophy of hard work, referring to a belief that hard work is the source of happiness in life; (4) reaction to automation, referring to a Pro or Con attitude toward automation and other technological innovations; (5) response to productivity increase, indicating whether the respondent regards productivity increase or industrial conflict as more important for improving the living conditions of workers; (6) opinion about labor–management cooperation, indicating whether the respondent considers that labor and management should cooperate or struggle with one another; (7) the respondent's support for the Socialist Party; and (8) orientation to traditional values, to be measured in terms of the respondent's approval of the "New Life Movement," a company-sponsored program to improve employees' family life, his preference for a superior who will look after subordinates' private lives, rather than a task-oriented superior, and his readiness to conform to others' views in order to avoid friction or conflict. For the last category, the respondent's affirmative responses were taken as an index of his inclinations toward "traditional" values. The questions designed to measure each of the eight categories are shown here only in abbreviated form. Some of these questions were omitted in the questionnaire for the Kokan Steel survey.

Table 3

Allegiance Types in Relation to Other Attitudinal Variables

Category and Question		A. Per Cent of Total Respondents Who Answered "Yes"	B. Per Cent of Each Allegiance Type Who Answered "Yes"				
			Pro- Pro	Pro- Con	Neut.- Neut.	Con- Pro	Con- Con
1. *Job Satisfaction*							
a. Do you like your present job?	Kokan[a]	44	63	44	39	28	22
	Shikoku[b]	46	69	52	42	49	34
	Deviation[c]		+ +	oo	oo	−o	− −
b. Is your job important for the company?	Kokan	72	82	68	67	69	68
	Shikoku	73	87	85	76	75	63
	Deviation		+ +	o+	oo	oo	o−
2. *Satisfaction with Workshop*							
a. Do your workshop members cooperate with each other?	Shikoku	68	87	73	75	73	49
	Deviation		+	o	o	o	−
b. Does your superior treat you squarely?	Shikoku	44	71	43	42	42	28
	Deviation		+	o	o	o	−
3. *Philosophy of Hard Work*							
If one works hard, should one be able to make a good living?	Shikoku	36	53	41	39	34	24
	Deviation		+	o	o	o	−
4. *Reaction to Automation*							
Do you support automation?	Shikoku	66	79	60	65	53	64
	Deviation		+	o	o	−	o
5. *Response to Productivity Increase*							
Do you think the first thing to be done for bettering workers'	Kokan	47	60	53	43	22	33
	Shikoku	65	86	68	75	60	45

Category and Question	A. Per Cent of Total Respondents Who Answered "Yes"	B. Per Cent of Each Allegiance Type Who Answered "Yes"				
		Pro-Pro	Pro Con	Neut.-Neut.	Con-Pro	Con-Con
living conditions is to raise the company's production level?	Deviation	++	oo	o+	−o	−−
6. *Opinion about Labor–Management Cooperation*						
Should labor and management cooperate with each other?	Kokan 64	74	68	63	43	52
	Shikoku 74	85	79	78	62	60
	Deviation	++	oo	oo	−−	−−
7. *Support for the Socialist Party*	Kokan 63	63	61	59	79	57
	Shikoku 61	55	58	56	75	66
	Deviation	oo	oo	oo	++	oo
8. *Orientation to Traditional Values*						
a. Do you approve of the New Life Movement?	Kokan 65	72	58	65	55	56
	Shikoku 47	54	48	45	61	43
	Deviation	++	−o	oo	−+	−−
b. Do you prefer a superior who looks after your private life?	Shikoku 38	34	50	40	31	37
	Deviation	o	+	o	−	o
c. Should one avoid conflict in life?	Shikoku 42	45	54	47	48	38
	Deviation	o	+	o	o	−

[a] The 1956 survey at Kokan Steel, all nine plants (N = 1,861).

[b] The 1957 survey at Shikoku Electric, whole company (N = 1,804).

[c] Of the two deviations under each listing, that on the left represents Kokan Steel, that on the right Shikoku Electric.

Figures below each column of the table indicate the percentages of the "yes" or "agree" responses to the questions. In Column A, the percentages of the "yes" answers given by the total respondents of Kokan Steel (all nine plants) and Shikoku Electric (the whole company) are shown, in order to indicate the direction and degree of deviation from this standard for each allegiance type. The percentages headed "Pro-Pro" in Column B signify the "yes" or "agree" response given by the Pro-Pro respondents only. This is followed in the same manner by the Pro-Con, Neutral-Neutral, and so on. The "Deviation" in each row indicates the direction of deviation for each type from the standard of the total respondents, by means of plus, zero, and minus signs, where plus means that the proportion significantly exceeded the standard, and so on. It will be seen from the table that differences do exist among the structures of allegiance types.

The Pro-Pro, or dual allegiance, type represents above all a person who holds a belief that hard work is always rewarding. Judging also from other responses that are not shown here, he is not the kind of person who tends to consider the causes of individual happiness or misery attributable to "society." This type of personality can be aptly characterized by "steady faithfulness." On the other hand, he is not necessarily a man with traditional value orientations. Furthermore, he is satisfied with his job as well as with his workshop group. It is indeed worth emphasizing that his satisfaction with these items is much higher than is that of the Pro-Con, or unilateral company allegiance, type. This may well be a surprise to many people, since the reverse is often taken for granted. Moreover, the dual allegiance type positively supports automation and productivity increases. He is also particularly favorably disposed to the idea of labor–management cooperation.

One is likely to associate the label "dual allegiance" with the image of a "smart yes-man" who wants to be favored by both union and management. True, there may be a small number of such people. However, according to a test specially designed to distinguish such superficial dual supporters from the workers with genuine dual allegiance, which we applied on the occasions of the more recent surveys, the proportion of the superficial supporters usually proved to reach only 10 per cent or so. A majority of Pro-Pro workers was found to be genuinely devoted to both union and company. Being faithful and earnest, they were active unionists, while at the same time worked hard on the job for the company.

One of the most salient characteristics of the Pro-Con, or unilateral company allegiance, type, on the other hand, is his conformity to traditional values. Particularly he favors the value represented by a *ninjo* (paternalistic) superior, who is ready to look after his subordinates' private lives. With respect to the other attitudes related to job, workshop group, automation, labor–management cooperation, and so forth, the Pro-Con type always scores zero, namely, his percentages being roughly equal to the standard. In other words, he is not dissatisfied, but, unlike the Pro-Pro type, he is not a positive supporter. At the same time, as the label "unilateral company allegiance" may connote, he is decidedly anti-union. When asked about possible union members' support for their union, those who agreed to the statement, "One need not conform to the decisions of the union as a whole, if one is not convinced of their appropriateness" were predominantly of the Pro-Con type.

As the characterization "nonpartisan" may suggest, the Neutral-Neutral type scores zero for all eight categories. In other words, he assumes an indifferent or lukewarm attitude toward most items. One cannot expect in workers of this type such eagerness and sincerity as are shown by the dual allegiance type.

The Con-Pro, or unilateral union allegiance, type, by contrast, shows clear-cut characteristics. What is most remarkable in this type is that he is an enthusiastic supporter of the Socialist Party. In Kokan Steel in particular, many workers of this type were supporters of the left-wing faction of the Party, which was well-known as a champion of class struggle.

It is interesting, in this connection, that the support for the right-wing faction of the Socialist Party was most numerous among the Pro-Pro workers at Kokan Steel. As can be imagined from the political inclination, the workers with unilateral union allegiance are skeptical about automation, critical of productivity increases, and tend to disagree with the idea of labor–management cooperation. Among the Con-Pro workers at Kokan Steel, for example, those who supported class struggle through the union movement amounted to 66 per cent, while those who agreed with the idea of productivity increases reached only 22 per cent. Likewise, in the same population, those who supported the ideology of labor–management conflict amounted to 53 per cent, whereas those who favored the idea of labor–management cooperation were limited to 43 per cent.

Finally, the Con-Con, or discontented, type shows deviations opposite to those of the dual allegiance type in almost all categories. A person of this type is extremely critical or suspicious of the philosophy of hard work. He is satisfied neither with his job nor with his workshop group. Like the Con-Pro, or unilateral union allegiance, type, he has negative attitudes toward productivity increases and labor–management cooperation. In the support of the Socialist Party, however, he is much less enthusiastic than the former. Again, it is this group that shows the minus deviation on the scale of traditional value orientations more clearly than any other type. In this respect, the discontented workers form a marked contrast to those with a Pro-Con attitude, namely, unilateral company allegiance.

In this connection, it should be noted that the discontented workers are not necessarily "disloyal" or "rebellious" to both union and management. At least two differing attitudes are distinguishable among those who outwardly show negative allegiance to both; the one may be considered a reflection of a rebellious or destructive frame of mind, apt to be against everything, while the other may be derived from especially high expectations and ideals, which naturally create a critical attitude toward the *status quo* in both union and company. Those critical of the *status quo* may well be basically loyal to both parties. According to a test designed to discriminate between these two subgroups of the Con-Con type, which I have made several times since the second Shikoku Electric survey in 1961, the destructive group always reaches less than 5 per cent, while 72 to 82 per cent of this category are those who maintain more or less constructive, though critical, attitudes. It appears that the discontented workers with higher educational attainment are likely to belong to the critical group.

The structural characteristics of allegiance types outlined above are not confined to the respondents of the two surveys at Kokan Steel and Shikoku Electric. Similar results have been observed in the data obtained from other Workers' Allegiance Surveys conducted elsewhere. When a similar method of analysis was applied to the data resulting from the survey conducted at Tokyo Electric Power in 1961, for example, the attitudinal characteristics of its employees in each of the five allegiance types were found to form exactly the same pattern as that shown in Table 3.

Worker Attributes in Relation to Allegiance Types

Having analyzed the structural differences among the five allegiance types, let us turn to the question of what inherent characteristics of workers are related to each. Table 4 outlines the relationships between allegiance types and a few important employee attributes, such as age, educational background, position in the company, and so forth. Here again the data shown in the table are those drawn from the 1956 Kokan Steel and the 1957 Shikoku Electric surveys. However, similar relationships between allegiance types and employee attributes were confirmed also by the data resulting from other Workers' Allegiance Surveys, including those at Kogaku Optics in 1955, Matsuya Department Store in 1961, Tokyo Electric Power in 1961, and Okamura Manufacturing in 1966.

In Table 4, figures indicate percentages by allegiance type for each of the respondent groups classified by age, education, and so on. For convenience of comparison, the corresponding percentages for the total respondents in the two companies are shown in the uppermost row.

As for age, the table reveals a considerable difference

Table 4

Worker Attributes in Relation to Allegiance Types

(Per Cent)

Worker Attribute		Pro-Pro	Pro-Con	Neut.-Neut.	Con-Pro	Con-Con	Others	Total
Kokan Steel[a]		32	6	8	7	9	38	100
Shikoku Electric[b]		15	7	10	5	27	36	100
Age								
Kokan	Less than 35 years	22	6	9	7	11	45	100
	35 years and over	38	6	7	5	6	38	100
Shikoku	20–29 years	12	6	8	6	35	33	100
	30–39 years	12	7	10	4	27	40	100
	40–49 years	23	6	14	6	21	30	100
Length of Service								
Kokan	Since 1945	25	6	8	6	11	44	100
	Since 1944 or before	44	5	5	5	3	38	100
Shikoku	Since 1951	12	5	9	3	37	34	100
	Since 1941–1950	12	7	10	6	29	36	100
	Since 1940 or before	22	7	12	5	19	35	100

Worker Attribute		Pro-Pro	Pro-Con	Neut.-Neut.	Con-Pro	Con-Con	Others	Total
Educational Career								
Kokan	Primary	35	6	7	5	5	42	100
	Lower Secondary	33	6	7	6	7	41	100
	Upper Secondary	22	6	9	7	12	44	100
	University	21	6	9	7	7	50	100
Shikoku	Primary	26	9	12	2	16	35	100
	Lower Secondary	19	10	12	6	21	32	100
	Upper Secondary	11	4	9	6	32	38	100
	University	10	4	10	4	37	35	100
Position in Company								
Kokan	Shop Operative	25	7	8	7	9	44	100
	Shop Supervisory	54	5	5	1	5	30	100
	Clerical	29	5	10	4	9	43	100
	Office Supervisory	29	5	8	6	4	48	100
Shikoku	Shop Operative	17	6	11	5	29	32	100
	Shop Supervisory	13	5	18	7	20	37	100
	Technical	6	2	11	3	40	38	100
	Clerical	11	5	10	5	32	37	100
	Administrative	23	11	11	4	15	36	100
Career as Union Official								
Kokan	Yes	37	4	9	9	5	36	100
	No	29	6	8	6	8	43	100
Shikoku	Yes	17	6	9	10	22	36	100
	No	15	7	11	3	29	35	100

[a] The 1956 survey at Kokan Steel, all nine plants (N = 1,861).
[b] The 1957 survey at Shikoku Electric, whole company (N = 1,804).

between above and below thirty-five. The Pro-Pro, or dual allegiance, type occupies a larger proportion of the older group, while among the younger, especially those in their twenties, the Con-Con, or discontented, type forms a significantly larger proportion, as compared with the company averages. Similar differences can be seen between those with greater and lesser length of service. For the allegiance types other than Pro-Pro and Con-Con, however, differences by age and length of service are not significant.

With respect to educational career, there is a general tendency for a group with lesser education to be dual in its allegiance.[5] Since the older age groups tend to have less formal education in postwar Japan, however, the larger proportion of the Pro-Pro type in the group with less formal education may be accounted for, at least in part, by age difference. On the other hand, the discontented type tends to form a larger share among those who have acquired higher formal education. As is shown in the table, the Con-Con type constitutes a larger proportion among university graduates in Shikoku Electric, and among upper secondary school graduates in both Shikoku Electric and Kokan Steel. No significant differences by educational background, however, are found in the other allegiance types.

Regarding position in the company, the dual allegiance type is significantly numerous among the shop supervisory group (54 per cent) at Kokan Steel, and among the administrative group (23 per cent) at Shikoku Electric. The latter tendency was confirmed also by the more recent surveys. It is to be noted, however, that the inclination to the Pro-Con, or unilateral company allegiance, is considerably stronger in the administrative group than in either operative or supervisory groups. On the other hand, the discontented type forms a larger share of engineering personnel (40 per cent) at Shikoku Electric. Though in a lesser degree, the discontented type also clusters more among clerical employees in both companies. As can be seen in Table 2 above, the dis-

5. Under the postwar Japanese educational system, formal education is given in the 6–3–3–4 years pattern, extending from primary school through lower and upper secondary schools to university, with the starting age at six. In this chain, compulsory education terminates with the completion of lower secondary school, normally at the age of fifteen.

contented type is especially conspicuous (43 per cent) among clerical workers at the Head Office of Shikoku Electric.

In this connection, an interesting result has been observed as regards the respondent's experience of the union office. One is tempted to expect the relative dominance of the Con-Pro, or unilateral union allegiance, type among those who have experience of union leadership positions, but this is not the case. If workers with experience as union officials and those without are compared, the inclination to unilateral union allegiance is much stronger among the former than the latter. However, either the dual allegiance or the discontented type is found to be most common among union leaders.

Finally, allegiance types also appear to be differentially distributed according to the nature of the jobs in which workers are engaged, as Table 5 indicates. The companies and plants classified under "Group A" in the table are equipped mainly with conventional types of machinery and rely heavily on physical labor, whereas those under "Group B" are technologically more advanced and demand more mental work which, in turn, requires workers with a higher formal education. The jobs and occupations in the former are mostly blue-collar jobs, while those in the latter are more like white-collar occupations. As is shown in Table 5, the dual allegiance workers predominate in Group A, whereas in Group B the discontented are more numerous and, in some cases, even exceed the dual allegiance employees. Moreover, among the Group B establishments, such automated plants as the new power installations at Tokyo Electric tend to produce a greater proportion of discontented workers than does the establishment as a whole.

The findings described above imply that differences in the nature of jobs and, accordingly, the underlying differences in equipment and system of work flow, strongly influence the distribution of allegiance types. We may therefore reasonably

Table 5

Distribution of Allegiance Types According to Work Process

(Per Cent)

Company and Plant	Pro-Pro	Pro-Con	Neut.-Neut.	Con-Pro	Con-Con	Others	Total
Group A							
Kokan Steel, Kawasaki Iron-works (1952)	34	7	9	3	10	37	100
Ube Industries, San'yo Colliery (1952)	42	2	9	9	2	36	100
Kokan Steel, Kawasaki Iron-works (1956)	37	8	9	4	6	36	100
Kokan Steel, Tsurumi Iron-works (1956)	34	2	7	11	5	40	100
Kokan Steel, Toyama Ironworks (1956)	46	1	5	15	2	31	100
Group B							
Kogaku Optics, Whole Company (1955)	22	6	10	8	16	38	100
Shikoku Electric, Whole Company (1957)	15	7	10	5	27	36	100
Kokan Steel, Mizue Ironworks (1960)	19	10	10	6	17	39	100
Matsuya Dept. Store, Whole Company (1961)	6	1	7	22	26	38	100
Tokyo Electric, Whole Company (1961)	28	5	14	6	12	35	100
Tokyo Electric, New Power Plants (1961)	11	4	20	10	28	27	100

expect that technological innovation now in progress will have considerable impact upon the development of industrial relations in the future. Judging from these research results, it appears probable that discontented employees will multiply, while the proportion of workers with dual allegiance will decline, as technological innovation proceeds further in many industrial establishments.

International Comparison of Worker Allegiance

One of the most striking characteristics of worker allegiance revealed by our surveys is that the positive correlation types, and the Pro-Pro and Con-Con in particular, tend to predominate, while the two negative correlation types, namely, the Pro-Con and Con-Pro, form so small a proportion that one may regard them as exceptional. In my opinion, this characteristic is not confined to the companies and plants studied. So long as a similar method of research is employed, the positive correlation types of allegiance will be found predominant in the majority of Japanese firms.

Scholars as well as intellectuals in general are apt to define the nature of industrial relations *a priori* as the relations between two diametrically opposing parties. With regard to workers' attitudes also, they tend to assume that company and union allegiance will be in negative correlation. Even I myself had a similar view at the outset of the surveys.

The same erroneous assumption is shared frequently by the leaders of both union and management. This can be demonstrated by the fact that these people are likely to assume that by weakening union members' attachment to the company they can by that much strengthen union solidarity, or that by repressing employees' concern for the union they can by the same degree heighten their company loyalty. In other words, they take it for granted that an inverse correlation between the two sets of allegiance is the rule. Yet, our empirical findings indicate that worker allegiance is neither unilateral nor "logical" as is often assumed by these people.

What, then, is the reason behind the widespread occurrence of a positive correlation between types of allegiance? Two reasons are conceivable. First, the fact that the "enterprise union" type of labor organization has been common in postwar Japan may account for this result.[6] Second, the reason may be found in the universal workers' ethos, which is characterized by their realistic and pragmatic attitudes toward their environment.

As for the first point, the great majority of Japanese local unions are intra-enterprise organizations composed solely of the employees actually working in a particular enterprise. They are of course not "company unions" in the American sense, for most of them have their sovereignty and can decide and act by themselves independently of management. Many of them are also affiliated with regional or national federations. Nevertheless, since it is an intra-enterprise organization, workers tend to perceive their union as an entity existent under the same roof as the company. In the eyes of workers, the company and their union are the two wheels of a cart, so to speak, and they should be complementay to each other, even though they may be in partial and temporary conflict. The prevalence of the dual allegiance type among Japanese workers may be explained in terms of this fact.

Even more important is the second reason, however. A series of worker attitude surveys conducted in the United States, as was mentioned earlier, also reports the predominance at least of the dual allegiance type. Father Purcell's studies provide a good example.[7] Based upon an assumption that the harmonious order of the "plant community" is more vital for workers than either their union or the company alone, Purcell hypothesized, at the beginning of his research,

6. As many as 94 per cent of Japanese local unions are of this type, according to recent statistics. For a discussion of the peculiar features of the "enterprise unions" in Japan, see, for example, Solomon B. Levine, *Industrial Relations in Postwar Japan* (Urbana: University of Illinois Press, 1958), pp. 90–91, 98–107.

7. Purcell, *Worker Speaks His Mind.*

that dual allegiance is the "normal state" of worker attitudes, and then proved—or at least he thinks he proved—this to be true. His research was, of course, different from mine both in the original hypotheses and methods, but there is an important similarity in the findings. Purcell found the dual allegiance type in 73 per cent of rank-and-file workers, in 57 per cent of the first-line supervisors, and in 88 per cent of shop stewards.[8] Though not as striking as Purcell's findings, the other studies also report similar tendencies.[9] It appears inadequate, therefore, to attribute the cause of dual allegiance solely to the particular characteristics of Japanese industrial relations.

The frequent occurrence of dual allegiance in the American arena may be explained by some characteristics peculiar to its industrial composition. For instance, the lack of bitter ideological struggles, or the "business-oriented" union activities which essentially accept the capitalistic economic system, may account for the phenomenon. To my way of thinking, however, it is more convincing to seek the cause in a more fundamental and more universal factor, namely, the worker ethos.

By "worker ethos" I mean a mental disposition to work honestly in the shop and, at the same time, to assert one's right to a fair share of the results of work. The ideal self-image under this disposition is the person who cooperates with management on the job as a good employee, while being ready to protect worker rights through the union, even by fighting against management.

Many industrial workers are neither idealists nor utopians,

8. Purcell published a second report of his study at the Swift Company in 1960. It is reported that in the East St. Louis plant, for instance, the dual allegiance workers amounted to 99 per cent of the total number of workers surveyed. This figure is incredibly high when compared with the findings of other surveys conducted in the United States, not to mention our findings in Japan. See Theodore V. Purcell, *Blue Collar Man: Patterns of Dual Allegiance in Industry* (Cambridge, Mass.: Harvard University Press, 1960), pp. 253–255.

9. See, for example, Stagner, *Psychology of Industrial Conflict*, pp. 400–403.

unlike some intellectuals. They do not perceive industrial relations as acute conflict relations between union and management as the leaders of the two parties often do. Their prime concern is not directed toward the drastic transformation of the existing socio-economic order, nor do they concentrate their energy in opposing the other party primarily on ideological grounds. Rather, they endeavor to secure their own livelihood through industrious work and, whenever possible, they attempt to improve their standards of living.

For many workers, their day-to-day lives and immediately obtainable interests are scored high on their value scales. This is their major criterion for evaluating company policies and union strategies. It seems natural, therefore, that such a realistic ethos should be shared more widely by those aged over thirty-five, who have family dependents, and who have acquired some responsible position within both company and union. To these people, the class-struggle theory which emphasizes irreconcilable conflict historically destined to occur between union and management, and which prescribes the destruction of the capitalist class as the prerequisite for worker welfare, sounds too idealistic. For them, the all-or-nothing type of logic which holds that the support of one party necessarily negates that of the other, or that a true unionist must be anti-company, appears too fastidious and is foreign to their daily way of living.

The high frequency of the Con-Con, or discontented, type among Japanese workers is also explicable in terms of the two reasons given above. From their viewpoint, both union and company should be jointly responsible, as it were, for improving workers' welfare and happiness. If the company is causing trouble and insecurity in workers' daily lives, the union is also held at least partly responsible for it. If, on the other hand, the union fails to protect workers' rights sufficiently, part of the responsibility is also attributable to the company, since in that case the company must have put pressure on the union to make it fail. In consequence, if one party is unable to fulfill its responsibility for workers, they are likely to criticize, or become negative in their attitudes toward, *both* parties, instead of only one. The occasional predominance of the Con-Con type workers can be accounted for in this way.

It would be a mistake, from the idealistic viewpoint of intellectuals, to regard such realities of worker ethos as their "backwardness" or "opportunism." In my opinion, the realistic orientation of workers discussed above reflects a sound wisdom which they have acquired through their daily lives and their work experiences. For the same reason, it is anticipated that the predominance of the positive correlation types will remain in the future, even if unions in Japan succeed in overcoming their structural limitation—namely, enterprise unionism—and metamorphose themselves into industry-wide organizations. Arnold Rose, for example, reports that workers with dual allegiance are common in his research of a teamster union. Many of the members of a Teamster Local at St. Louis working for fifteen different firms are reported to be loyal to both union and company at the same time.[10] The existence of a strong industry-wide union organization will not necessarily negate the workers' union and management allegiance.

Concluding Remarks: Toward Constructive Industrial Relations

The foregoing analyses of worker allegiance provide us with suggestive clues as to what we should—or should not—do in order to develop constructive industrial relations. Be-

10. Rose, *Union Solidarity*, pp. 65–70.

fore discussing such policy implications, however, a few preliminary comments may be necessary.

In my opinion, constructiveness and democracy are the two important qualifications for the kind of industrial relations which can be called "modern" or "modernized." To modernize industrial relations of a country, such as Japan, therefore, it is vital for the leaders of both unions and companies to exert conscious efforts to make them democratic and constructive.

By "democratic" industrial relations I mean the following: first, they are the kind of union–management relations where both parties engage in bargaining on an equal footing and in full recognition of the rights of each other. Second, industrial relations may be called "democratic" when sufficient opportunities are given for the employees of a company to participate in its managerial decisions. In other words, "democratic" means "equal" and "participative."

Of these two, the first hardly needs explanation. With respect to participativeness, I shall express my view in some detail in Chapter Six of this volume, in connection with the question of employee participation in management as a prerequisite to democratic industrial relations.[11] In the following, therefore, I shall discuss mainly the other qualification for modernized industrial relations, namely, constructiveness. It is obvious that this quality is primarily concerned with the organizational relations between union and company.

"Constructive" industrial relations are by no means synonymous with free-and-easy union–management cooperation. As was discussed earlier, the organizational aspect of industrial relations is essentially characterized as one of opposition. Though the opposition may be a latent one, it is quite possible for it to develop into open strife between union and management. Paradoxically, relationships cannot be truly constructive unless both parties are prepared to engage in a

11. See Chapter Six, section on "The Development of Industrial Democracy."

battle if necessary. Relations will become constructive when a strong union gets together with a strong management, both seeking a mutually workable solution to the issues of conflicting interests through collective bargaining. It is most important in such a confrontation that each of the two parties pay its respect to the rights and *raison d' être* of its opponent.

It is at this point that my research findings on worker allegiance seem suggestive. Generally speaking, company executives tend to believe that the Pro-Con, or unilateral company allegiance, type is the only group truly loyal to management. In the eyes of many executives, this type is the stabilizing factor in industrial relations. On the basis of this belief, management often attempts to cultivate Pro-Con workers through many—some direct, some indirect—methods. The traditional, oppressive and despotic approaches having been rendered obsolete, the more recent strategies in this respect include various manipulative techniques under the disguise of employee welfare services, "human relations" programs, or specially arranged educational activities such as the "New Life Movement." By means of tactful measures of this kind, management sometimes seeks to convert not only the Con-Pro type, but also the Pro-Pro type, into the Pro-Con.

Such efforts, unfortunately, are doomed to failure. For the union, however undesirable and defective its activities may seem to management, is the indispensable cornerstone for the lives of the working people. It is impossible to deprive a worker of his role as a union member. This may be demonstrated by the fact that very few workers of Pro-Con type have been found in our surveys.

Moreover, such a fruitless effort is likely to invite unfortunate consequences. If management mounts an anti-union campaign in its eagerness to develop unilateral company allegiance, it will be a threat to the union leaders, who will in turn accelerate their efforts to cultivate unilateral union allegiance among workers. Again, if management tries to minimize the *raison d' être* of the union, to estrange union officials from rank-and-file members, or to conciliate union leaders, such strategies may result, quite unexpectedly, in transforming the employees to the Con-Pro, or even Con-Con, type. For they are likely to regard such anti-union tactics as undue pressure by management upon the union, and this critical attitude, in its turn, will induce them to be critical of the union itself. What management should do instead is try to increase the proportion of dual allegiance workers, by converting the Con-Pro, Pro-Con, and particularly the Con-Con, types into the Pro-Pro. Management should realize that dual allegiance, not unilateral company allegiance, is the stabilizing and constructive factor in industrial relations.

Union leaders, on the other hand, often assume that only the unilateral union allegiance type represents the authentic trade unionists. As a result, many union leaders are eager to transform rank-and-file members into this particular type, by eliminating their loyalty to management. In fact, I was often asked by the union representatives, when meeting with them to interpret the survey results, what they should do to convert workers with dual allegiance to the unilateral union loyalty type. This concern seems to have been enhanced in recent years in connection with their effort to overcome enterprise unionism.

The union leaders' attempt deliberately to turn workers into unilateral union men, however, will not bring about the anticipated results. Where else could a worker earn his livelihood and find a place to work outside his own company? In other words, any union member cannot be totally indifferent to his role as an employee; even the most elaborate worker education program sponsored by the union cannot possibly change this fact. Only in a few exceptional cases, where workers have a particular reason to be proud of or grateful to the union, but resentful of the existing policies of management, will a large proportion of the workers spontaneously convert their allegiance into the unilateral union type.[12]

Furthermore, this kind of effort on the part of union leaders may have unexpected consequences. For example, the strategies that give an excessive emphasis to class struggle —behaving as if it were the eve of a revolution, denouncing management as the "enemy" in class warfare, and so on— will not only invite a stiffened attitude on the part of company representatives, but eventually may alienate the leaders themselves from rank-and-file members. Union splits, which have been by no means rare in postwar Japan, are likely to result from such excesses. This was the case, for example, in the serious union splits which took place some years ago, in the midst of a bitter, long struggle at Oji Paper and Pulp and at the Miike Colliery of Mitsui Coal Mining. Without doubt, deficient management policies must first shoulder the blame in these cases. However, union leaders' misguided efforts to force the workers to take on unilateral union loyalty were also a factor in these splits.

This leads me to my concluding remarks. The leaders of both unions and companies have often competed in trying to force-develop workers with their respective unilateral types of allegiance. Not much good has come out of this competition, but only instability—or even destructiveness— for industrial relations. In fact, the excessive competition to create either the unilateral union or the unilateral company allegiance workers has aggravated the inherently bitter union-management relations in Japan. In order to cultivate constructive industrial relations, therefore, the leaders of both parties must give up their desire for the unilateral

12. Out of the many companies and plants studied, only three such exceptional cases have so far been experienced. See Tables 2 and 7.

Table 6

Workers' Allegiance Surveys Conducted by University of Tokyo, 1952–67

Year	Company	Sample Size	Proportion to Universe	Research Technique	With Collaboration of
1952	Kokan Steel Kawasaki Ironworks	701	$\frac{1}{19}$	Interview	Company Personnel Department and Union Executives
1952	Ube Industries San'yo Colliery	336	$\frac{1}{8}$	Interview	Company Personnel Department and Union Executives
1955	Kogaku Optics Whole Company	1,831	Total	Questionnaire	Company Personnel Department, Union Executives, and Rikkyo University
1956	Kokan Steel Nine Plants	3,555	$\frac{1}{6}$	Questionnaire	Company Personnel Department, Union Executives, and Institute of Statistical Mathematics
1957	Shikoku Electric Whole Company	1,854	$\frac{1}{3}$	Questionnaire	Company Personnel Department, Union Executives, and Institute of Statistical Mathematics
1959	Isetan Department Store Male Employees	556	Total	Questionnaire	Company Personnel Department and Union Executives
1960	Kokan Steel Mizue Ironworks	1,285	Total	Questionnaire and Interview	Company Personnel Department and Union Executives

Year	Company	Sample Size	Proportion to Universe	Research Technique	With Collaboration of
1961	Shikoku Electric Whole Company	2,726	$\frac{1}{2}$	Questionnaire	Company Personnel Department, Union Executives, and Institute of Statistical Mathematics
1961	Matsuya Department Store Whole Company	2,420	Total	Questionnaire and Interview	Company Personnel Department and Union Executives
1961	Tokyo Electric Whole Company	5,537	$\frac{1}{5}$	Questionnaire	Company Personnel Department, Union Executives, and Institute of Statistical Mathematics
1963	Kokan Steel Whole Company	5,547	$\frac{1}{8}$	Questionnaire	Company Personnel Department, Union Executives, and Rikkyo University
1965	Shikoku Electric Whole Company	2,696	$\frac{1}{2}$	Questionnaire	Company Personnel Department, Union Executives, and Kagawa University
1966	Okamura Manufacturing Whole Company	606	Total	Questionnaire	Company Personnel Department, Union Executives, and Rikkyo University
1967	Jeco Watchmaking Whole Company	1,044	Total	Questionnaire	Company Personnel Department, Union Executives, and Rikkyo University

allegiance of workers. They should instead have a better understanding of the basic worker ethos and try to encourage dual allegiance among workers. Again, an earnest effort should be made to transform workers of discontented inclination into those with dual allegiance, by providing them with the objective conditions under which they can be diligent employees and ardent unionists at the same time. If only the leaders of both parties appreciate this requirement, it will help industrial relations to become more constructive. And I believe that such a need is not limited to the Japanese industrial setting alone. Judging from the fact that the leaders of both parties in Western countries also tend to contend with each other for the loyalty of workers, the need to develop dual allegiance among workers must be emphasized there also, in order to keep their industrial relations constructive.

Table 7

Distribution of Allegiance Types in Seven Companies Other Than Kokan Steel and Shikoku Electric

(Per Cent)[a]

Company	A. Company and Union Allegiance					B. Combined Allegiance Types						Total	Number of Respondents	Distribution Pattern
	Pro-M	Pro-U	Con-M	Con-U	Inclination	Pro-Pro	Pro-Con	Neut-Neut	Con-Pro	Con-Con	Others			
Ube Industries (1952) San'yo Colliery	52	72	16	9	U	42	2	9	9	2	36	100	(336)	X
Kogaku Optics (1955) Whole Company	37	43	34	28	U	22	6	10	8	16	38	100	(1,603)	X
Isetan Department Store (1959) Male Employees	28	36	29	21	U	15	3	22	6	12	42	100	(549)	Y
Matsuya Department Store (1961) Whole Company	10	41	67	31	U	6	1	7	22	26	38	100	(2,155)	Y
Tokyo Electric (1961) Whole Company	42	45	25	25	–	28	5	14	6	12	35	100	(5,178)	X
Okamura Manufacturing (1966) Whole Company	45	29	27	47	M	21	11	9	3	21	35	100	(562)	X/Z
Jeco Watchmaking (1967) Whole Company	8	49	75	24	U	7	0.3	4	31	22	36	100	(1,011)	Y

[a] Percentages show the frequency of occurrence to the sample size, excluding the "don't know" and blank responses.

CHAPTER FIVE

Workers Think of Their
Work and Leisure

Company Allegiance and Work Motivation

Workers' attitudes toward their work are generally considered to be correlated to their attitudes toward their work place, and consequently toward the company and plant where they are employed. According to this view, the stronger their allegiance to the company, the greater will be their work motivation. As we shall see below, however, these two attitudes are not always positively correlated. In fact, there are cases in which the degree of workers' company allegiance is high but their work morale is low, and vice versa.

In the case of Japanese workers, company allegiance has in general declined in recent years, and particularly since the latter half of the 1950s. Since the boom which started in 1956, workers' wages have continued to rise, their working hours have been steadily reduced, and the equipment and welfare facilities have been markedly improved. Despite these improvements, however, it appears that the attitude of unquestioning loyalty to their company and management, which was common among workers in prewar Japan, is now gradually disappearing.

128

This tendency is particularly conspicuous in those firms which employ a larger number of non-manual or white-collar personnel. As I have shown in the previous chapter, the employee attitude surveys I have conducted since 1952 reveal that in those companies where manual or blue-collar jobs are more numerous, workers with a favorable attitude toward the company tend to outnumber those with unfavorable or critical opinions. The reverse is, however, true in those companies whose business involves more non-manual, white-collar work. More important, where automation has been introduced, there is a tendency for more of the employees to adopt an attitude of opposition toward the company.[1]

This last phenomenon may be accounted for by the higher level of education of those in automated jobs, and by the fact that such jobs not infrequently consist of monotonous, repetitive labor. The gradual decline in workers' allegiance to their company, however, appears to be a general trend in recent Japan, not confined to these specific types of firms. In my four surveys conducted at Kokan Steel, for example, I found that each successive study showed a decline in Pro attitudes toward the company and an increase in Con attitudes, as will be seen in Table 8.[2]

In view of this recent general trend for Japanese workers, the level of education and the tedium of work alone provide only insufficient grounds for explanation. In other words, it seems appropriate to recognize a general decline in the sense of belonging to the organization where one is employed, and a gradual obsolescence of the practice of "lifetime commitment" to it, a commitment which has been fairly common in large-scale enterprises in Japan until recently. In place of these traditional orientations, it seems that self-

1. See Chapter Four, section on "Worker Attributes in Relation to Allegiance Types."
2. See Chapter Four, section on "The Distribution of Allegiance Types."

Table 8

Company Allegiance at Kokan Steel, 1952–63

(Per Cent)[a]

Year	Pro-Company	Con-Company	Others	Total	Number of Respondents
1952: Kawasaki Ironworks	54	18	28	100	(701)
1956: Nine Plants[b]	50	21	29	100	(1,861)
1960: Mizue Ironworks	43	28	29	100	(1,051)
1963: Whole Company[c]	26	44	30	100	(3,917)

[a] Percentages show the frequency of occurrence to the sample size, excluding the "don't know" and blank responses.

[b] Include Kawasaki and four other ironworks, one brickfield, and three dockyards.

[c] Includes Mizue and five other ironworks, three dockyards, and the head office.

consciousness as free individuals and a spirit of independence have at last begun to develop among Japanese workers. A fact that reveals this change at the behavioral level is the increase in labor mobility between enterprises since the latter half of the 1950s.[3]

How, then, do Japanese workers think about their work? In my surveys cited above, I asked the respondents the following question: "Of all the knowledge and skills necessary for your future life, which do you think is the most important?" In 1963 at Kokan Steel, 66 per cent of the workers replied to this question, "I wish to acquire knowledge and skills related to my job." This proportion is overwhelmingly larger than that of workers who replied that they wanted to acquire general education and culture, such skills as flower arrangement or driving, or to learn more about the company. When the same question was put in 1965 at Shikoku Elec-

3. The rate of separation by regular workers in all manufacturing industries, for example, was only 23 per cent in 1957, but was 25 per cent in 1960, 28 per cent in 1963, and 31 per cent in 1964. Particularly conspicuous was the increase in the number of those who left for personal reasons, which amounted in 1964 to close to 90 per cent of the total separated. Rodo-sho (Ministry of Labor), *Rodo hakusho* (Labor white paper; Tokyo: Rodo-sho, 1966).

tric, a firm with many non-manual workers, as many as 81 per cent of the sample returned the same reply. This we may take as indicating the employees' deep concern for their own work.

With regard to the kind of work desired, the first preference was everywhere for a "job in which I can make the fullest use of my abilities," though next to this was the wish for one from which a high income could be earned. At Okamura Manufacturing, one of the largest makers of steel furniture in Japan, these two answers alone were given by 66 per cent of the respondents. Similar preferences were found when the same question was asked of a sample of ordinary citizens, in a national survey on social stratification and social mobility carried out in 1965, where the sum of these two preferences came to approximately 55 per cent.

As can be seen from the above, the desire to extend one's abilities and to excel in one's occupation, and at the same time to possess economic power—in short, the desire for worldly success—are important motives for work. This fact was affirmed by workers' views of the reasons for success. At Okamura Manufacturing, for instance, the first factor considered essential for success was "effort" and the second "talent," the sum of the two replies amounting to 67 per cent of the sample.

It is only natural for people with such a view of work to believe that only those with real ability should be promoted and, eventually, attain success in life. When asked, therefore, whether opportunities for promotion in the company should be decided by workers' length of service and school career, or alternatively by their real ability, 71 per cent of the employees at Shikoku Electric (1965), 76 per cent at Kokan Steel (1963), and as many as 84 per cent at Okamura answered that the decisions should be related mainly, or at least give due consideration, to ability. This tendency is very closely related to workers' age and educational background.

At Okamura, for example, those who stressed ability were particularly numerous among young workers between twenty-five and thirty-four (91 per cent), and among the graduates of upper secondary schools and universities (87–90 per cent).

Job Satisfaction

From the foregoing it seems clear that, despite a general decline in allegiance to their company, the workers' motivation for work itself, at least potentially, is fairly high. The question then is whether or not they are able fully to realize their inclination to work, and to exert their talents and abilities to the full. The existing state of things does not seem to be very promising in this respect.

For example, at a number of firms I put the question: "Do you think that your present job is important for the company?" The answers are shown in Table 9. In all cases more than 56 per cent of the workers answered "Yes." Thus they had a high evaluation of their work, and some pride in it. However, when the same workers were asked further if they felt that their work, which was so important, was "appreciated by the company," their replies were rather pessimistic. No more than 37 per cent felt that it was properly appreciated. The negative answers, "I don't feel that it is appreciated" and "I don't know whether it is appreciated or not," amounted to approximately 70 per cent. If we define those workers who feel their work is important but not properly appreciated by the company as "alienated," the proportions in this category were about 37 per cent at both Kokan Steel and Okamura, and 45 per cent at Shikoku Electric.

Many workers also felt that the company's policy regarding promotions and wage increases was "unfair" because,

Table 9

Workers' Evaluation of Their Work in Five Companies

(Per Cent)

Company	Important	Average	Not Important	Total	Appreciated	Don't Know	Not Appreciated	Total	Number of Respondents
Matsuya Department Store (1961)	56	36	7	100	21	47	31	100	(2,420)
Tokyo Electric (1961)	79	17	3	100	37	36	26	100	(5,537)
Kokan Steel (1963)	68	27	5	100	33	42	25	100	(5,547)
Shikoku Electric (1965)	70	25	5	100	26	47	26	100	(2,696)
Okamura Manufacturing (1966)	61	35	4	100	27	53	19	100	(606)

133

while these decisions should give primary consideration to workers' real abilities, management still attached too great an importance to the traditional factors of seniority and educational background. No more than 10 per cent of the employees in the companies cited above felt that management policy was "fair."

Predictably, job satisfaction among these workers was not great. When asked if they liked their present jobs at the company, only about 40 per cent replied in the affirmative, as will be seen in Table 10. In other words, more than half the employees were to some degree dissatisfied with their jobs.

True, those who had higher positions in the company tended to be better satisfied. At Kokan Steel, for instance, while only 36 per cent of the rank-and-file workers expressed

Table 10

Workers' Satisfaction with Their Work in Five Companies

(Per Cent)

Company	Satisfied	Undecided	Dis-satisfied	Total
Matsuya Department Store (1961)	34	48	17	100
Tokyo Electric (1961)	41	46	13	100
Kokan Steel (1963)	36	49	14	100
Shikoku Electric (1965)	45	45	10	100
Okamura Manufacturing (1966)	40	48	10	100

satisfaction, 65 per cent of the section heads were content. However, even fewer rank-and-file employees of these companies than ordinary citizens were in fact satisfied with their jobs, as our Stratification and Mobility Survey in 1965 showed that 57 per cent of the latter were contented with their work.

If we compare the proportion of those satisfied with their jobs in Japan with that in various countries of the West, we find that the Japanese group is considerably smaller. This point is made clear by a comparison with the data on job satisfaction in six countries furnished by Alex Inkeles.[4] As can be seen in Table 11, only West German manual workers, with 21 to 47 per cent satisfied with their jobs, roughly correspond to the Japanese in this respect (approximately 40 per cent). The degree of job satisfaction among Japanese workers does not even come up to that of Italian manual workers. It does not come anywhere near the figures for manual workers in the United States, Sweden, and Norway.

What can be the reasons for the higher proportion of workers satisfied with their jobs in such countries as Norway, Sweden, and the United States? This is a question which requires careful study. One may argue that the workers are more satisfied in these countries because personnel management there is more skillful than in Japan. At least one of the

Table 11

National Comparisons of Job Satisfaction[a]

(Percentage Satisfied)

Type of Job	U.S.A.	West Germany	Italy	U.S.S.R.[b]		Sweden	Norway
White Collar	82	33–65	–	60	Middle	72	88
Skilled Manual	84	47	68	62	Class		
Semiskilled Manual	76	21	62	45	Working	69	83
Unskilled Manual	72	11	57	23	Class		

[a] Adapted from table 1 in Alex Inkeles, "Industrial Man," *American Journal of Sociology,* 66 (July 1960), p. 6.

[b] Data for U.S.S.R. are those of Soviet refugees provided by the Russian Research Center, Harvard University.

4. Alex Inkeles, "Industrial Man: The Relation of Status to Experience, Perception, and Value," *American Journal of Sociology,* 66 (July 1960), pp. 5–8.

reasons may be the comparative freedom workers in these countries enjoy in their activities in the work place. Further, the fact that workers in these countries find it comparatively easy to transfer from one company to another may also have contributed to their greater job satisfaction.

Certainly the ideal pattern of employment relations is one in which the objective conditions are provided for movement of workers from one company to another without any great disadvantage, and at the same time where the workers themselves are enthusiastically attached to their present job and, consequently, to their place of work. Although the state of employment relations in Japan has recently shown an increase in the rate of labor mobility, it is still far from this ideal.

Patterns of Leisure Activity

How, then, do the workers of modern Japan, who hold views about work like those described above, spend their leisure hours?

In prewar Japan, it was a privilege confined to the "leisure class" in Veblen's sense to enjoy leisure fully and openly.[5] The working masses of those days, of course, had some leisure time. However, it was spent in most cases in rest or took the form of humble diversions at home. For the majority of working people, pleasures pursued outside the home —for example, excursions, visiting the geisha quarters, theater-going, or hunting—were rare, simply because they were very expensive. Besides, since working hours were much longer than now, people had little time to spare for such pleasures. More important, however, the prevailing code

5. Thorstein B. Veblen, *The Theory of the Leisure Class: An Economic Study in the Evolution of Institutions* (New York: B. W. Huebsch, 1899).

of values before the war regarded such pleasures as "corrupting."

This code was based on the Confucian ideology of the Edo period. Confucianism, the philosophy of conduct of life officially sponsored by the shogunate, regarded the enjoyment of leisure as dangerous to the established order. This idea was expressed in such maxims as "Pleasure seeking is a bar to diligence, and in the long run will ruin one."

The influence of the philosophy of Ninomiya Sontoku (1787–1856), the well-known thinker, should not be overlooked in this connection. Sontoku, who succeeded in rehabilitating many impoverished agricultural villages in the Kanto district toward the end of the Edo period, preached thrift and diligence. "Work comes first" was the cardinal point in his teaching. He vigorously exhorted people to work even harder and to maximize the production of goods providing food, clothing, and shelter. Behavior which did not live up to this high standard was censured as "idleness." Another central item of his teaching was the idea that the articles produced by hard work should be economized as much as possible, and that one should not consume in a manner disproportionate to, or beyond, one's means or status. Behavior contrary to this was rejected as "extravagance."

It should be noted that Sontoku's philosophy consisted not alone in praise of thrift and diligence. It also taught that only those who followed these teachings and worked hard would attain success, while those who violated them and gave first place to pleasure, or consumed in a manner unfitting to their status, would necessarily become failures in life. His thought, which exerted a great influence on the view of the art of living held by the common people of Japan, from the time of the Meiji Restoration up to the Second World War, was a "philosophy of success" similar to that of Benjamin Franklin.

It was because of this established code of values that before the war the enjoyment of leisure was frequently criticized, and despised, as idleness or extravagance. In fact, it was not infrequent in those days for people, even when they had the chance, to enjoy their leisure as inconspicuously as possible.

This pattern of leisure activities of the Japanese masses, and the values underlying them, however, were suddenly swept away after the end of the Second World War. This change was of course not peculiar to Japan, but was part of a world-wide change brought about by the arrival of the "era of leisure." Its arrival in Japan became conspicuous after the economic boom starting in 1956. Especially after the early 1960s, the number of people openly enjoying leisure has increased markedly year by year. At the same time, a new way of thinking which holds that it is only right and proper to enjoy one's leisure during off hours, on weekends and holi-days, is rapidly becoming widespread. As the causes of this new situation, known in the Japanese language as *reja bumu* (leisure boom), we may cite the following points.

First of all, there has been a general rise in the income level of the working masses. Since the end of the 1950s the lower stratum with an annual household income of less than two thousand dollars has decreased rapidly, while the middle-income stratum with between two and three thousand has increased considerably. In 1964 the latter was 2.8 times as large as it had been in 1960.[6] No doubt such incomes do not seem sufficient to permit any great enjoyment of leisure, when compared with those of Western workers. In fact, many workers in Japan complain about the scantiness of their wages. Nevertheless, it is undeniable that the recent increase in income has made it possible for them to spend more money than before on leisure.

The recent reduction of working hours in many enter-prises, which has resulted mainly from technological inno-

6. Keizai Kikaku-cho (Economic Planning Agency), *Kokumin seikatsu hakusho* (White paper on national living conditions; Tokyo: Keizai Kikaku-cho, 1965).

vation, has also contributed to the "leisure boom." In contrast to an average 60-hour week before the war, the hours worked in all manufacturing industries had been reduced to 50 by the end of the 1950s. More recently, there has been a decline in the number of companies employing even the 48-hour working week, and an increase in firms adopting a 45- or 42-hour schedule. As a result, the average weekly leisure hours in 1960 as compared with 1941 had increased from 23.8 to 36.4 hours for manual workers, and from 32.2 to 37.1 hours for white-collar workers.[7]

Further, as a result of mass production, plenty of consumer durables and a variety of instant and frozen foods are now available, and their prices have been reduced, so that families are now able to make a marked saving in the time and energy hitherto required for housework.

An increase in the facilities for leisure is another cause. Since the latter half of the 1950s, there has been a considerable growth in the number of such amusement places as pinball halls, bicycle racing tracks, baseball grounds, dance halls, cabarets, bowling alleys, golf courses, skiing grounds, and so on. In addition, highways and expressways required for motoring and excursions are steadily under construction.

More important, however, is the change in people's values with regard to leisure. This change may be characterized as a transition from a way of thinking which places work at the center of life to a view which regards leisure as the aim of living.

The flood of leisure activities unleashed by these factors has been spreading year by year, and some intellectual leaders have come to feel obliged to give warnings to the masses, or to criticize the new pattern of values which places leisure at the center of life.

There are, however, several kinds of activities which today go by the name of "leisure." Provisionally, we may dis-

7. *Ibid.*, 1964.

tinguish four categories. The first is a kind of leisure which takes the form of mere rest or is pursued only for the sake of maintaining one's health. The term "recreation" seems to have originally referred to such activities. For our present purposes I shall call this category "rest leisure."

The second comprises a variety of leisure activities pursued only as a spectator. This includes watching television, listening to music, reading newspapers and books, seeing films, attending baseball games, and so on. I shall call this kind "spectator leisure." Some of these pastimes of course may also provide rest and recreation. A characteristic common to both categories is that their behavior is "passive."

In contrast to the two above, the third involves "active" behavior in which one actually performs. I shall call this "participant leisure." Hiking, motoring, golfing, skiing, photography, gardening, raising animals, stamp collecting, and the like, fall into this category. Also to be included here is the practice of various arts and skills, such as taking lessons in piano, flower arrangement, or dressmaking, as well as holding periodic reading parties, study groups, and so forth. All the activities often called "hobbies" are of the participant type.

In addition, there are such activities as going to church, attending P.T.A. meetings, participating in various community projects, and engaging in public welfare services, which may be considered a type of leisure. Do-it-yourself activities on holidays may also be classified as of this genre. Some sociologists, calling these activities "semileisure," distinguish them from leisure pursuits in the proper sense, on the ground that they also fulfill some sort of obligation.[8] In the sense, however, that they do not aim at any material re-

8. Joffre Dumazedier and Nicole Latouche, "Work and Leisure in French Sociology," *Industrial Relations,* 1 (February 1962), p. 21; Dumazedier, *Toward a Society of Leisure,* trans. Stewart E. McClure (New York: Free Press, 1967), p. 93.

ward, as does one's occupation, these activities can be regarded as leisure in a broad sense. This fourth category I shall provisionally call "service leisure."

Turning now to the actual frequency with which the different kinds of leisure activities are engaged in by the Japanese, most common until recently have been those of the spectator type. According to a survey conducted in 1960–1961 by the Japan Broadcasting Corporation's Cultural Research Institute, 50 to 60 per cent of the Japanese spent their leisure hours after 6 P.M. on weekdays in watching television or listening to the radio. Next came mere relaxation by reading newspapers or weekly magazines and by "lying down dozing," two forms of resting, about 30 per cent of them spending the above hours in these activities. By contrast, only about 10 per cent employed the same hours for leisure activities of either the participant or the service type. As regards the length of time spent at leisure, the national average on a weekday amounted to 4 hours and 50 minutes, of which almost 3 hours were spent in either resting or spectator leisure.

The time allocation on holidays was of course a little different. When activities at 10 A.M. and 2 P.M. were examined, it was found that the proportion using these hours for television and the radio diminished to about 20 per cent, while those spending them in outings, sports, hobbies, or in visiting friends increased greatly. Among the younger generation, for example, 60 to 70 per cent used these hours in participant leisure activities. On the whole, the time employed in participation increased approximately three times, and the total leisure time amounted to 7 hours and 12 minutes for the national average on holidays.[9]

Similar patterns of leisure activities have been observed among the workers at several companies where I have con-

9. Nihon Hoso Kyokai Hoso Bunka Kenkyusho (Japan Broadcasting Corporation Cultural Research Institute), *Nihonjin no seikatsu jikan* (The time budget of the Japanese; Tokyo: Nihon Hoso Shuppan Kyokai, 1963), p. 253.

ducted studies. Take, for example, the case of Kokan Steel in 1963. When asked how they spend their leisure hours throughout a week, a majority, or 56 per cent, of rank-and-file workers answered that they spent them on television, radio, newspapers or magazines. Ranked next were "seeing films, plays, or sports activities" and "gardening or raising pets," but neither accounted for much beyond 10 per cent of the respondents. Only 6 per cent of the workers answered, "taking part in sports myself." In Okamura Manufacturing, on the other hand, 13 per cent of the employees answered, "taking part in sports." As in Kokan Steel, however, the greatest number of workers cited "television, radio, newspapers or magazines," though the proportion was smaller (36 per cent).

It is evident that leisure activities of the Japanese, at least until recently, have been mainly of the spectator and resting types. Television watching, in particular, has been by far the most common among their leisure pursuits. Strictly comparable data are not available from other developed countries. In a survey conducted in the United States in 1954, some 1,500 respondents were asked how they would spend the time if they could have two extra hours per day; it was found that only 4 per cent answered that they would watch television, while 25 to 40 per cent would "relax," and 11 to 28 per cent would "spend it in reading or in study."[10] It is clear, however, that these data are not comparable with those cited above. According to another study made in the 1950s, most Americans watched their television sets on the average of 18 hours a week.[11] This is a very high figure, if the average

10. Alfred C. Clarke, "Leisure and Occupational Prestige," in *Mass Leisure,* ed. Eric Larrabee and Rolf Meyersohn (Glencoe, Ill.: Free Press, 1958), pp. 211–212.

11. Rolf Meyersohn, "Social Research in Television," in *Mass Culture: The Popular Arts in America,* ed. Bernard Rosenberg and David Manning White (Glencoe, Ill.: Free Press, 1957), pp. 345–357.

weekly leisure hours at that time for the Americans are esti-
mated at 36 to 38 hours.[12] Again, it is reported, based upon
inquiries made in 1959, that in France, most people watched
television on the average of 16 hours a week, or a little over
2 hours a day. In the evening particularly, the majority ate
before looking at television, and at 8 P.M. the percentage of
viewers amounted to 75 per cent.[13] From these facts one may
conclude that even in Western countries, the spectator type
of leisure activities has been predominant.

The situation is rather different in Soviet Russia. According to a survey conducted in Leningrad in 1961, 22 per cent
of the weekly leisure hours were spent by workers in what is
called "learning," which includes attending adult education
courses, receiving political education, and self-study, while
62 per cent were used for "rest and amusement," which comprised most leisure activities of the resting and spectator
types. Again, studies conducted at two enterprises in Kazan
in 1963 and at several factories in Alma-Ata in 1964 revealed
that, of some 32 leisure hours a week given to the employees,
only 20 to 30 per cent were spent in television watching and
seeing films or plays.[14]

It must be noted, however, that more recently a tendency
has developed for the Japanese to spend more time in participant leisure. When the official statistics for 1960 and 1964
are compared, those who actually spent leisure time in travel-
ing, hiking, or motoring had increased to 11 per cent by the

12. According to a study conducted in the 1950s, the average hours spent
daily in leisure by Americans, holidays and weekdays inclusive, were 5 hours and
8 minutes for blue-collar males and 5 hours and 24 minutes for white-collar
males. See George A. Lundberg et al., "The Amount and Uses of Leisure," in
Mass Leisure, ed. Larrabee and Meyersohn, pp. 178–180.

13. Dumazedier, *Society of Leisure*, pp. 155–156.

14. Hiromi Teratani, "Yoka" (Leisure), in *Gendai Soviet shakai ron:
Shakaigaku-teki bunseki* (Contemporary Soviet society: A sociological analysis),
ed. Akira Tsujimura (Tokyo: Nihon Kokusai Mondai Kenkyusho, 1970), pp.
114–137.

latter date, as compared with only 5 per cent earlier. Even more important, the proportion who in 1964 said that they would have liked to engage in these activities was as large as 42 per cent. Likewise, the proportion who cited some kind of sport played by themselves increased from 4 per cent in 1960 to 13 per cent in 1964.[15]

This new tendency is also evidenced in terms of the change in expenses people make for their leisure activities. Official statistics show that while there was a considerable growth in per capita expenditures for consumer durables between 1957 and 1960, there has been a rapid increase since 1960 in expenditures for leisure activities. During the period from 1960 to 1963, for example, the average rate of increase for all households in Japan of leisure expenditures amounted, by item, to between 16 and 19 per cent. Especially conspicuous was the growth in amounts spent for travel, mountain climbing, hiking, motoring, and skiing, resulting in a "travel boom."[16]

Finally, when a comparison is made by age group, it is clear that there is a tendency for the younger generation to prefer the participant to the spectator type, while older groups tend to spend more time in resting or in visiting friends and other social activities. Analysis by income group, on the other hand, shows that the resting type is most common among those of low income, whereas those who mention do-it-yourself activities, sports, travel, and other forms of participant leisure are more commonly of the middle- and higher-income strata.[17]

15. Economic Planning Agency, *White Paper on National Living Conditions,* 1965.
16. *Ibid.*
17. *Ibid.,* 1961, 1962, and 1964.

Work and Leisure: Five Types of Worker Attitude

In the foregoing we have outlined the main features of leisure activities in modern Japan. Let us now turn to the workers' attitudes toward their leisure. Do they still think that work, rather than leisure, is the center of life? Are they, instead, inclined to be leisure-oriented, as some intellectual leaders have feared?

In this section, workers' attitudes toward their place of work, their income level, and their work itself will be examined. In other words, we put to ourselves the following questions: How do they regard their home life in comparison with their work at the company? Which is more important to them, an increase in income or increased chances of leisure? What do you think about the work–leisure dichotomy?

With regard to the first question, I have already cited data from five different companies in my Workers' Allegiance Surveys. One of the questions commonly used in these surveys asked the workers to choose, out of a list of items, the one which they felt contributed most to making their life worth living. The list included such items as "work at the company," "leisure spent in hobbies and amusements," and "making and keeping a happy home." Their answers are shown in Table 12. As may be surmised from the decline in company allegiance discussed in a previous section, a very small percentage of the employees in each company (6 to 14 per cent) replied, "I feel that my work at the company makes life worth living." By contrast, a much larger proportion (25 to 47 per cent) felt that "leisure" made life worth living, and to my surprise, the highest proportion (36 to 56 per cent) felt that "making and keeping a happy home" was the center of their lives.

Table 12

Workers' Preference among Work, Leisure, and Home

(Per Cent)

Company	Work	Leisure	Home	Others and Unknown	Total
Matsuya Department Store (1961)	6	47	36	11	100
Tokyo Electric (1961)	7	32	56	6	100
Kokan Steel (1963)	11	25	51	13	100
Shikoku Electric (1965)	8	30	55	6	100
Okamura Manufacturing (1966)	14	28	38	21	100

Another point in this table which deserves note is that at Matsuya Department Store, where the majority of employees (58 per cent) were young females, a higher figure was obtained for "leisure" than for "making a happy home," while at Tokyo Electric and Kokan Steel, where female employees were very few (6 and 4 per cent, respectively), the opposite was the case. Similar findings showing the leisure-oriented attitudes of young females, however, have been obtained from our other surveys.

At any rate, what is really at issue here is a choice between the work place and the home, rather than between work and leisure. When asked more directly which they thought to be more important, the work place or the home, the workers tended to choose the latter. In Okamura, for example, 32 per cent chose the home, while only 15 per cent preferred the work place. It is interesting, however, that more workers (41 per cent) answered, "Both the work place and the home are important."

For those who work in industrial establishments, the choice between income and leisure is closely related to that between wage increases and reductions in working hours. It is reported that in some Western countries, in Sweden, for example, the majority of workers desire more leisure time even at the expense of reduced earnings. According to our findings in Japan, however, at least at four firms studied, the great majority (about 80 per cent) replied, "Wage increases should come before a reduction in working hours." This suggests that Japanese workers still have not attained an economic power sufficient to enable them to enjoy their leisure. There are exceptions to this rule, however. When the same question was put to the workers at a certain petrochemical industry plant in 1964, 43 per cent were in favor of reduced working hours, as compared to 36 per cent in favor of higher wages.

What, then, do the workers of modern Japan think about the work–leisure dichotomy? Obviously, this is the very core of the question we have been concerned with. Before we look into this point, however, let us first classify the possible ways of thinking about this dichotomy. What is at issue here is not a simple either-or choice between the absolutely work-oriented and the absolutely leisure-oriented. Apart from these two extreme cases, we may expect that there will be different views of living, depending upon whether one regards work and leisure separately or whether one integrates the two.

First, as regards the either-or choice between work and leisure, we may consider the way of thinking of Ninomiya Sontoku as representing one extreme, in which hard work is the basic principle of life. This type of attitude, which I shall call "work-oriented," may be summed up in the following statement which our respondents were asked to evaluate: "Work is man's duty. I want to devote myself wholly to my work without any thought of leisure."

At the opposite pole, there is a way of thinking which holds that it is leisure that makes life really worth living. Bertrand Russell once advocated this view in his article, "In

Praise of Idleness."[18] This attitude, which I shall call "leisure-oriented," may be represented by the following statement presented to my respondents: "Work is no more than a means for living. The enjoyment of leisure is what makes human life worth while. I want to enjoy leisure fully."

It goes without saying that these two views are incompatible. Besides these two, however, people may take approaches which incorporate both work and leisure, and which recognize some significance in each. The distinction here is whether one merely segregates work and leisure within the framework of one's daily life, keeping the two in a state of mutual separation, or whether one links them in a coordinated way so that the two will be in a state of mutual stimulation. If the former approach is chosen, life will be seen as dualistic and will be divided into two unrelated fields. This can make life easier and more comfortable than the latter approach, which requires a strong will and much more energy. A statement characterizing the former, which I shall call the "split type," was offered in my surveys as follows: "Work is work and leisure is leisure. Modern man gets his work done smartly, and enjoys his leisure reasonably."

If, on the other hand, one is to take the latter approach and to integrate work and leisure into a condition of unity, continuous coordination of the two will be needed. In other words, what is required here is to maintain a stance in which, by devoting oneself to one's work, one will be better prepared to enjoy leisure without reserve, and by so doing will be all the more disposed to devote oneself to work. This attitude, which I shall call the "integrated type," was described to my respondents in the following terms: "Hard work makes leisure really enjoyable, and the full enjoyment of leisure gives new energy to work. I want to work to the best of my ability, and to enjoy leisure fully."

18. Bertrand Russell, "In Praise of Idleness," in *Mass Leisure,* ed. Larrabee and Meyersohn, pp. 96–105.

Besides these four, there is a fifth possible point of view which sees work as one's principal pleasure and so is itself a form of leisure, even of play. This approach, which I shall call the "identity type," was expressed in my surveys in the following statement: "My only pleasure is work. There is no distinction between work and leisure. I therefore need not be liberated from work in order to enjoy leisure."

From the above we may distinguish five different types of attitude with regard to the relations between work and leisure. They are: A, work-oriented; B, leisure-oriented; C, identity; D, split; and E, integrated.

In my recent surveys conducted at Okamura Manufacturing in 1966 and at Jeco Watchmaking in 1967, statements designed to express the five above patterns were shown to the respondents, and they were asked to select that closest to their real attitude. In addition, in another survey carried out in 1966 by the Institute of Statistical Mathematics, the same questions were asked of a sample of ordinary residents in the twenty-three wards of Tokyo. The numbers of respondents in these three surveys were 606 for Okamura, 1,044 for Jeco, and about 800 for the Tokyo residents. The results are shown in Table 13.

In the data presented in Table 13, the facts which attract our attention are: that the proportion of those who chose the leisure-oriented (B) type was in each case unexpectedly

Table 13

People's Preference among Five Attitudes Related to Work and Leisure

(Per Cent)

Company	A Work-oriented	B Leisure-oriented	C Identity	D Split	E Integrated	Others and Unknown	Total
Okamura Manufacturing (1966)	12	5	7	23	51	2	100
Tokyo Residents (1966)	8	6	6	22	51	7	100
Jeco Watchmaking (1967)	4	7	5	20	64	–	100

small; that the majority always chose the split (D) or integrated (E) type; and that more than half of the respondents in each case regarded the integrated type as representing their own attitude toward life.

A survey containing somewhat similar questions was carried out by the Institute of Journalism at the University of Tokyo in 1959, taking a sample of some 900 ordinary citizens of Tokyo. Using a set of four, instead of five, statements concerning the relationship between work and leisure as questions, the survey elicited answers fairly like those obtained in my surveys. The four statements were as follows: "Since work is a duty of man, one must work as hard as possible," a statement which will be referred to here as the "A prime" because of its resemblance to my work-oriented type; "Work is a form of pleasure. I have never thought it necessary to be liberated from work so that I might enjoy myself," which will be referred to as the "C prime," since it is a slight modification of my identity type; "I like work, but I wish to have time sufficient for recreation and amusements by means of which I can supply energy for work," which will be called the "D prime"; and "Work is work and leisure is leisure. I will work hard during the stipulated hours, but when I am free I want to forget all about the work and enjoy leisure," which I shall call the "E prime." The proportions preferring each of these four were: 19 per cent for A prime, 12 per cent C prime, 21 per cent D prime, and 39 per cent E prime, with 9 per cent for "others and unknown."[19]

By cross-tabulating the results from my Okamura study with the workers' age, education, and position in the com-

19. Tokyo Daigaku Shinbun Kenkyusho (University of Tokyo Institute of Journalism), "Tokyo tomin no seikatsu jikan to seikatsu ishiki" (The time budget and attitudes toward life of the citizens of Tokyo), *Shinbun Kenkyusho Kiyo,* 10, no. 2 (1961). "Others" include in this survey the proportion who chose as their own attitude the statement: "Work is nothing but a means to earn one's living," which amounted to 4 per cent.

pany, I was able to obtain the following findings, as shown in Table 14. Among older age groups, and particularly those in their forties, an appreciable proportion thought of the work-oriented (A) type as representing their own attitude toward life. The identity (C) type was also picked out more frequently by those above fifty than by younger workers, excepting teenagers. By contrast, among workers in their twenties, those who preferred the integrated (E) type were conspicuously numerous. As regards educational background,

Table 14

Worker Attributes and Five Attitudes Related to Work and Leisure

(Per Cent)

Worker Attribute	A Work-oriented	B Leisure-oriented	C Identity	D Split	E Integrated	Others and Unknown	Total
Okamura Manufacturing (1966) Whole Company	12	5	7	23	51	2	100
Sex and Age							
Male							
Less than 19 years	2	12	11	29	47	–	100
20–29 years	5	5	3	23	62	2	100
30–39 years	21	3	7	23	45	1	100
40–49 years	24	4	8	16	47	1	100
50 years and over	21	5	13	18	41	3	100
Female							
Less than 29 years	4	4	2	26	63	2	100
30 years and over	15	5	12	22	42	5	100
Educational Career							
Primary and New Lower Secondary	14	4	10	25	45	2	100
Old Middle	26	7	16	11	40	–	100
New Upper Secondary	5	6	4	26	60	–	100
Old Higher and University	10	4	3	20	63	–	100
Position in Company							
Temporary Operative	15	12	15	18	38	3	100
Regular Operative	8	5	8	25	53	1	100
Supervisory	18	3	5	24	49	2	100
Technical	23	7	7	7	55	2	100
Administrative	22	4	4	22	48	–	100

those who chose the integrated type were found in much larger numbers among those with a high level of education, and especially those who graduated from schools at the postwar upper secondary or higher levels. Finally, with respect to position in the company, those who preferred the work-oriented type were most numerous among those who held either technical or higher positions in the company.

Concluding Remarks: Summary and Prospects

As far as can be determined from the survey data quoted above, the work motivation of modern Japanese workers is fairly high, despite a general decline in allegiance to the company and management.

This positive inclination to work itself, however, has not been fully utilized, nor have the workers been able to exert their real abilities to the fullest extent. The failure here seems due largely to defects in managerial leadership and organizational structure. As a result, the workers' talents tend to be obscured within the work place and neglected by management, and even by themselves. The fact that more workers are dissatisfied with their jobs than in the West is a consequence.

Under such circumstances it is only natural that Japanese workers, and particularly the younger ones, try to compensate for their job dissatisfaction by the enjoyment of leisure. The question here is whether they have gone too far in this direction, so as to have become absolutely leisure-oriented and utterly indifferent to their work and to the work place. As our surveys show, however, the absolutely leisure-oriented attitude is by no means common among Japanese workers. It is of course true that one of the reasons for the relative infrequency of this attitude is the workers' financial weakness, which has so far not allowed them extensive opportunities to enjoy leisure. If their economic conditions improve greatly, people preferring this orientation may multiply.[20] Even if they do, however, they will not exceed those who prefer the integrated position.

The attitude supported, and actually adopted, by the majority of younger workers is neither leisure-oriented nor of the split type. Rather, it is of the integrated type, a pattern in which one's devotion to work and the full enjoyment of leisure are combined in a single process and rhythm of life.

On the other hand, older workers are still fairly work-minded, and the proportion preferring either the work-oriented or the identity posture is by no means small.

The work–leisure dichotomy commonly found in modern Japanese society, therefore, arises between the absolutely work-oriented and the integrated approaches, rather than between those that are absolutely either work- or leisure-oriented. Since both of the preferred choices are work-minded, we may conclude that the working masses of Japan today are still industrious and ambitious, regardless of age.

The question for the future is therefore whether younger workers will be able to find jobs to which they can devote themselves or work places where they can fulfill themselves. The introduction of a system of worker participation in management, which I have introduced in Chapter Two and shall discuss fully in Chapter Six, is a basic step to the solution of this problem.

20. In my survey at Okamura, I asked the respondents to give their *opinions,* namely, the stances they thought they should take, in addition to their actual attitudes, already reported. Their reactions were fairly different from those shown in Table 13. Especially, the proportion supporting the leisure-oriented (B) type reached 24 per cent, instead of the mere 5 per cent in the table. This suggests that if their financial conditions do improve, the percentage adopting this approach will appreciably increase, absorbing some portions of the respondents now taking the split or integrated attitude. Even when asked their opinions, however, the proportion showing the integrated attitude was greatest. The reactions of the Okamura respondents to the questions about their opinions were: 25 per cent for A, 24 per cent B, 8 per cent C, 6 per cent D, and 35 per cent E.

Another point which requires consideration is that workers must be provided with sufficient time and income, so that they not only can afford the leisure they desire, but also can fit their leisure orientations into their more basic work-oriented approach to life. What is important in this connection, however, is that the working masses should know how to integrate the full enjoyment of leisure and the devotion to work into a single process of life, and how they can make the integration more productive and creative. Without this knowledge, the enjoyment of leisure may result only in the waste of time and energy. It will also be necessary for them to acquire new life skills and social training, by which they can utilize their leisure for the enhancement of their culture and, therefore, for their human development.

CHAPTER SIX

A Program for Workers'
Participation in Management

Introduction

A New Year's address delivered by the president of a firm to encourage his workers to more strenuous efforts seems to be much the same everywhere. In Japan today, one frequently finds such a statement as the following in these addresses: "This year is going to be a crucial one for our company. You people are earnestly requested, in cooperation with us in management, to try your best to bring about better results for the company."

Strangely enough, however, one very rarely finds in such an address any mention as to how the workers are expected to do their best in their daily work. For example, they are never told under what concrete plans or standards they are expected to work. It is almost as if they were expected to make blind and random efforts at company goals. Perhaps some particularly enthusiastic men will indeed rush into work on their own initiative. The majority, however, will remain unchanged in their customarily lazy attitudes toward work, hardly ruffled by the urgency felt by the president and executives of the company. The New Year's address is thus likely to turn out to be no more than an annual routine.

Moreover, the workers are provided with very little chance to contribute any original ideas, criticisms, or desires they may have to the process of planning and formulating the programs and standards under which they are to work. It is certainly very rare that a New Year's address contains an appeal for the workers to participate in the making of decisions about ways and means to achieve company goals even in an "exceptionally crucial" year. Without participation, however, the workers will never consider the ways and means they use to be their own, nor will they fully understand why they are needed. As a result, they are again required to make blind and, even if enthusiastic, pointless efforts.

In recent years it has been customary for Japanese businessmen to talk with emphasis about "respect for humanity" or "promoting original ideas" among employees. It is never clearly stated, however, how employees can develop their humanity in their work, or how they can apply their originality to their job. Management seems to assume that humanity and originality develop automatically.

There are of course those company presidents who urge their employees to "speak up" whenever they are dissatisfied with life within the workshop. While one may admire the progressive spirit of such invitations, one wonders how the employees are to speak up if there is no institutional means of doing so. Without recognized procedures, the workers are likely to feel it rather risky to speak up freely.

In short, it is not enough to demand of workers more effort, more endurance, or more spontaneity. What is really necessary is to show them exactly why such strenuous efforts are required, and how they can be realized in the daily activities of the workshop. Most important, however, is that workers be encouraged to participate in the process of planning and of deciding the ways and criteria by which they are to work for company goals, and that their participation be institutionalized in the formal organization of the company. For several years I have discussed the vital need for a system which I call "workers' participation in management."[1] I have not, however, had a chance to describe the historical necessity for their participation, or the steps by which it is to be institutionalized in the formal organization of a firm. In the following, I shall discuss these points in some detail.

Workers' Participation in Management

To begin with, I shall make it clear what I mean by "workers' participation in management."

1. The term, as used here, refers primarily to participation by the employees who actually work in an enterprise. Although this kind of action is compatible with another form of worker participation, namely, by the labor union, these two should not be confused.

2. At the present stage of development of industrial democracy, employees have the right to participate in managerial decisions. It is, however, management that recognizes this right and provides employees with actual chances to take part.

3. Employees may participate in the decision-making processes at the top and middle levels, as well as in the shop, or bottom, level of an enterprise. The nature of decisions in which the employees participate, therefore, will be different at each level.

4. Employee participation should be put into operation as

1. See, for example, Kunio Odaka, "Sangyo no kindaika to keiei no minshuka" (Modernization of industry and democratization of management), *Chuo Koron,* no. 884 (July 1961), pp. 26–44. Chapter One of the present volume includes a modified version of this article.

a formally recognized system; in other words, it should be institutionalized in the organization of the enterprise as a whole.

5. Although the system will have a positive effect on the employees' work morale and therefore on their productivity, it should not be used manipulatively by management.

6. To bring the system properly into operation, a few preparatory steps will be necessary; for example, gaining employees' support, training them for participative practices, placing the right persons in the right positions, and so on.

7. If the system is successfully installed, it will give the employees ample chance to exercise their full talents and abilities, as well as an awareness that they are their own bosses at work.

8. Employee participation is the basis for democratizing industrial management. Only through a proper introduction of the system into enterprises will industrial democracy, at its present stage, be realized.

To amplify these points: 1. "Workers' participation in management" might comprise any of three different forms, namely, participation in management by employees, or by labor unions, or workers' self-management. Of these, the third form, in which the "workers' council," a body of worker representatives, holds the supreme power to control and manage the industrial undertaking as a whole, will be possible only under a general social framework where "social ownership" of the means of production is legally secured.[2] In its ideal form, there will be no managerial prerogatives, nor will management be a self-recruited class distinct from that of the workers. By definition, however, I presuppose in "workers' participation" the existence of managers who can

2. At present, probably Yugoslavia is the only country where this kind of workers' self-management is formally established under state legislation. See International Labour Office, *Workers' Management in Yugoslavia* (Geneva: International Labour Office, 1962); Jiri Kolaja, *Workers' Councils: The Yugoslav Experience* (London: Tavistock Publications, 1965).

have their own right to make independent decisions, and this third form may therefore be excluded from the present discussion.

"Union participation," on the other hand, may be of at least two different types: co-determination by union and management, in which case the two parties decide jointly on major managerial policies and share responsibility for carrying out their decisions; and joint consultation between union and management, where both parties confer to reach formal agreements on those matters in which they have common interests. Of these, only the first can properly be termed "union participation."

It is doubtful, however, whether a union should perform such a role, so long as it wants to retain its identity. To my way of thinking, co-determination will deprive a union of its unique social function as an "opponent" of management. If a union were to be directly involved in the planning and execution of managerial policies, it would be held at least partly responsible for their outcome. Among other things, it would never be able to go on strike over an issue which resulted from a co-determined decision. The German system of "co-determination" (*Mitbestimmung*), which is often referred to in this connection, is basically a form of employee participation. A majority of those representing labor on the supervisory board (*Aufsichtsrat*) are employees actually working in the firm. Although many of them are chosen by the German Federation of Labor Unions and the industrial union involved, they are not necessarily union officials.

Joint consultation is the only form of "union participation" that will not jeopardize the essential role of unionism. A union can remain an "opponent," even while consulting with management on such matters as technological innovations or the improvement of safety and sanitary facilities, where both parties have common interests. Along with collective bargaining, which originally was their major function, this kind of activity will become more important for unions in the future.

In joint consultation, however, it is the union, not the individual employee, which is the constituent. Attendance at joint consultative meetings is always confined to union officials. There is thus a sharp distinction between joint consultation and employee participation, where all the workers of an enterprise are involved, though in my opinion the two systems can well coexist.

2. The reasons why employees, at the present stage of industrial democracy, have the right to participate will be discussed later. The point at issue here is that their right must be recognized and given effect by management. As was stated above, employee participation presupposes the existence of managers, whose major role is to control and direct the operation of the physical and human organization of a firm as a whole. They have the prerogative to manage, chiefly because they are responsible to the public for the efficient, as well as the effective, operation of the organization. It is, in my opinion, also one of their prime responsibilities that they recognize the employees' right to participate and encourage them to do so.

3. People are apt to interpret the words "participation in management" to mean primarily sharing in managerial decisions at the top level of an enterprise, such as those bearing on long-range production plans, technological innovations, new plant construction, and the like. My concept, however, is that participation should be of a broader scope, ranging from top to bottom. Particularly, I attach importance to participation in the making and carrying out of decisions relating to the work of those at the bottom, or the shop, level of a firm. The opportunities of participation should be first given to this level, because employees here are in greater need of them than are those at higher levels.

Generally speaking, the higher the position, the greater

chance do employees have to participate. At the top level in particular, most employees are at the same time management. By contrast, the majority of workers at the bottom echelons of a company are merely the managed, who are expected to work always under the direction of others. The democratization of industry, in my opinion, should be first undertaken at the lowest levels of the enterprise.

However, the nature and scope of the decisions in which rank-and-file workers are to take part will naturally be confined to matters within the workshop. It will be impractical, if not impossible, to expect a shop-level operator to participate in the decision as to whether a new automated plant should be erected, or how much money should be borrowed for it from a bank. Such decisions will have to be made at the top level, with possible participation by those in middle management. Shop-level decisions, on the other hand, will be confined to such matters as, for example, the establishment of a time schedule for production goals that have provisionally been set at top and middle levels, the improvement of methods the shop-level operators are expected to follow in their daily work, the reform of supervisory practices, or the improvement of communication within a workshop.

4. Employee participation should be institutionalized into the formal organization of the enterprise as a whole. In other words, it should not remain only an informal practice, or be confined to only certain parts of the company. Participative leadership that has been informally initiated by a section head may be all right if it is of provisional form, but certainly not if it is a full-scale system of employee participation. The "human relations" techniques of improving communication or of "democratizing" supervisory practices within a workshop should also be considered as only a provisional step.

5. There are ample research findings to show that partici-

pation has a positive effect on employees' morale and their motivation to work.[3] The question here is whether participation should be used merely as a convenient device for motivating employees to work. As will be seen later, the beneficial effects we can gain from employee participation are not limited to this. Besides, frankly speaking, if it produces no such effect at all, I suspect that the average manager will never give due consideration to its institutionalization. Despite this benefit, however, it should be emphasized that employee participation is more than a convenient device. It is, as I said before, a right of employees to cooperate with management. Regardless of his status and his personal relations with top managers, every employee is a partner with them and should be treated as such. Moreover, today many top managers are themselves employees. As Peter F. Drucker points out, contemporary society is an "employee society," and in the United States in particular, this stage began as early as the 1930s.[4] In present-day Japan also, the "owner-managers" or the "capitalists" who were predominant in prewar years, have been largely replaced by high-ranking employees customarily called "salaried directors" or "professional managers," who in most cases run the large-scale enterprises. As employees, their formal status is not different from that of even shop-level operators, and it is only their functions as managers that vary. Since this similarity is the

3. The results obtained from the famous "Relay Assembly Test Room" of the Hawthorne Experiments are a classic example. Similar effects of such kinds of participation have been corroborated by a number of social psychological studies made by the University of Michigan Survey Research Center. See, for example, Lester Coch and John R. P. French, Jr., "Overcoming Resistance to Change," *Human Relations,* 1 (August 1948), pp. 512–532; Nancy C. Morse and Everett Reimer, "The Experimental Change of a Major Organizational Variable," *Journal of Abnormal and Social Psychology,* 52 (January 1956), pp. 120–129; John R. P. French, Jr. et al., "Employee Participation in a Program of Industrial Change," *Personnel,* 35 (November–December 1958), pp. 16–29.

4. Peter F. Drucker, "The Employee Society," *American Journal of Sociology,* 58 (January 1953), pp. 358–363.

necessary outcome of a historical development, it is only proper that a manager should treat a common employee as his equal and partner. Correspondingly, an ordinary employee has the right to participate in management.

6. With regard to the steps by which employee participation is to be institutionalized, I shall describe them in more detail later. A word, however, may be added here about the training of employees. Some business executives in Japan are doubtful or hesitant about the introduction of a system of employee participation. One of the reasons they give for their negative attitude is that employees themselves are generally lacking in the ability to participate and the willingness to do so. This may, to a certain extent, be true. However, this means that executives are all the more obliged to launch the needed employee training if participation is to be institutionalized. To my way of thinking, an enterprise should not remain only an establishment for production or for making profits. It should also be a training school where the workers learn a new way of thinking and acting, as well as how to exert their talents and abilities as fully as possible.

7. Employee participation, when successfully introduced, will have several beneficial effects for employees. These benefits, however, will not accrue only to the workers. The enterprise itself and management as well will share. Even the labor union will find the effects of employee participation positive. Despite their general resistance, labor unions will not be damaged or find their regular activities infringed upon. These points, however, will be discussed at length in the last section.

8. There are a good many definitions of democracy. In my opinion, however, democracy is most appropriately defined in terms of the extent to which members of a group can participate in the making and the carrying out of decisions of primary concern to the group and its members. In other

words, the degree of democratization of a group can be measured by the scope and depth to which its members are able to share in such decisions and their execution. It goes without saying that any group must have a leader or leaders who are authorized to make final decisions about operations and management. Sometimes these leaders may have to make decisions against the will of certain members. Nevertheless, if the degree of participation has been sufficiently high, even such decisions can be considered to be autonomous and made by the membership as a whole. The first requisite for democracy is not necessarily that the leaders of a group are fairly elected by its members. Rather, it is the extent to which these members can freely contribute their own ideas, desires, and criticism, as well as their own efforts, to the operation and management of the group.

The Development of Industrial Democracy

The term "industrial democracy" generally refers to the democratization of industrial relations, and particularly of the relations between labor unions and management. There is, however, another aspect of industrial relations, namely, employer–employee relations. The ultimate goal of industrial democracy, in my opinion, is to democratize fully the latter aspect of industrial relations in a given society.

Union–management relations, by nature, are those between two organizations which, under the control of different leaders, have different aims and functions. A labor union is basically an organization outside the jurisdiction of management. This holds true even with regard to the Japanese labor unions. As was pointed out earlier, the great majority of local unions in Japan are enterprise unions—that is, organizations composed almost entirely of the employees working at a single enterprise.[5] One may almost call them "intra-enterprise organizations." Nevertheless, they are not company unions and are basically independent of the managerial jurisdiction. It is for this reason that Japanese local unions are able to restrict the actions of managers by putting public pressure on managerial prerogatives through collective bargaining and disputes.

Individual employees, on the other hand, are of course working for a company under the control of managers, though they at the same time can be members of a local union. They are thus under dual control—that of management representing one organization, and that of union leaders representing the other.

The two aspects of industrial relations—union–management relations and employer–employee relations—are in reality closely inter-connected. Factors determining one are often found interlocked with those determining the other. Nevertheless, the forms of democracy in these two aspects are not necessarily alike. While democracy in employer–employee relations is primarily characterized by employee participation, union–management democracy consists in the union's right and ability to negotiate or confer with management as an equal body. Where unions have this right, they are also able to affect managerial decisions and to restrict the arbitrariness of management. This effect which a union can have upon management may be called "participative," since participation includes those actions that in some way or other affect the managerial decision-making process. In this sense, and in this sense alone, there is a similarity between these two forms of democracy in industrial relations.

5. See Chapter Four, section on "International Comparison of Worker Allegiance." According to recent statistics, more than 90 per cent of local unions in Japan are of this type.

How, historically, has industrial democracy been developed in these two aspects of industrial relations? In the following, I shall briefly outline this process.

The development of industrial democracy first takes place through the democratization of employer–employee relations. In brief, this is the shift from industrial slavery to industrial democracy. The type of industrial relations appearing at the *first* stage of development was the relationship between "employees" as slaves, servants, or retainers, and the "employer" as their master or lord. In the West, this type was predominant from ancient to medieval times. The noted German economic historian Werner Sombart, in his typology of industrial relations, cites the following examples for the ancient and medieval periods: that of house slaves and a patriarch in the early Greek periods; that of serfs and their masters in the Roman Republican era; that of dependent artisans and a city-dwelling plutocrat who was their owner in the later Roman periods; that of patrimonial tenant farmers and a landlord in medieval Europe; and that of apprentices or journeymen and their masters in the handicraft industry of the medieval European cities.[6] Of these, the first four examples roughly correspond to our type of industrial relations in the first stage.

Characteristic of this type is that the relationship of master and servant was primarily that of ownership, rather than employment. Servants here were "owned" by their master or lord, not merely as a labor force, but as human beings. In this relationship, however, the owner was not necessarily despotic, even cruel, nor were the servants absolutely or inhumanly subordinated. According to Sombart, most industrial relations in those days were based on what he calls the "solidarity" principle, that is, they were essentially mutualistic.

6. Werner Sombart, "Arbeiter," in *Handwörterbuch der Soziologie,* ed. Alfred Vierkandt (Stuttgart: Ferdinand Enke, 1931), pp. 7–12.

Industrial relations typical of the *second* stage of the development were those between "employees" as family workers, apprentices, or followers, and the "employer" as their head or boss. Sombart's example of apprentice–master relationship corresponds well to this type. Similar examples, however, may be found even today. Among the small-sized enterprises of Japan, this sort of relationship still persists in abundance.

Characteristic of this type is that it was a paternalistic relationship between a boss and his following. The boss here was at one and the same time the owner, manager, and a worker in the business, while his followers either were his family members or were treated as if they were. Sombart considers this kind of relationship typical of industrial relations based on the "solidarity" principle.

The *third* stage of the development roughly corresponds to the years from the industrial revolution up to the 1930s. Industrial relations in this stage may be called "capitalistic." "Employees" here were generally considered to be laborers and wage earners, in short, manpower, while the "employer" was a capitalist, an entrepreneur, or an owner–manager. This type of relation was of so impersonal a character that it was often conceived as an abstract relation between labor and capital. Employees were usually depersonalized, made into nothing but units of labor, which in their turn were treated as "commodities." The relationship was a seemingly rational one, of give-and-take between employees who were selling their labor in order to make a living, and employers who were buying it, in exchange for an equivalent.

At the same time, there was a new industrial despotism underlying this type of industrial relations. Employees were formally free men and were considered equal to their employer before the law. Actually, however, they were little more than "wage slaves" in that they were obligated to work under whatever working conditions were imposed upon them by the employer. Under the latter's despotic control, employees could not possibly be their own bosses in work, and therefore they were likely to lose their will to work. They felt they were alienated from their work. They also had very little chance to get out of the status of wage earner, and therefore tended to bear a class hatred toward the employer as a capitalist.

Industrial relations in this stage were thus characterized by two apparently unrelated factors: on the one hand, they were rational, an exchange of commodities for money; on the other, they were impersonal and exploitative.

Without doubt, the characterization I have given is somewhat exaggerated. However, I am now describing what Max Weber calls "ideal types," and some emphasis on certain elements of reality is inevitable.[7] Among the examples given by Sombart as representing capitalistic industrial relations, the following may be cited: the relationship of factory workers and capitalists in England in the first half of the nineteenth century; that of industrial workers and entrepreneurs in Germany in the 1930s; and that of wage earners and industrialists in the United States in the same period. According to Sombart, all these examples are based on the egoistic, exploitative, "individualistic" principle.[8]

Industrial relations in the *fourth* stage of development is the type which today, and particularly after World War II, has gradually appeared in advanced industrial countries all over the world. By "advanced industrial countries" I mean not only those in Europe and North America, but the Soviet Union and Japan as well. In other areas, perhaps this new

7. An "ideal type" is not an idealized form or model constituted from a certain value standpoint. It is a concept constructed by deliberately emphasizing certain factors or aspects of reality. It is generally considered a useful tool for making a historical or cross-cultural comparison of social phenomena. See Max Weber, "Die Objektivität sozialwissenschaftlicher und sozialpolitischer Erkenntnis," in *Gesammelte Aufsätze zur Wissenschaftslehre* (Tübingen: J. C. B. Mohr, 1922), pp. 190–205.
8. Sombart, "Arbeiter."

type of relation has not yet emerged; and even in the advanced countries, in terms of sheer numbers of examples, the third type, rather than the fourth, may still constitute the majority. Nevertheless, it is undeniable that this new type is increasing in numbers and that this process cannot be reversed now.

What characterizes this new type is that it is a relationship between a group of *employees* in a managerial capacity, and another group, also of *employees,* who are cooperating with the former as their partners. Both sides of the relationship, in other words, are represented only by employees. They are the same in formal status, being differentiated only in their functions. Moreover, there are a number of employees between these two, who have to varying degrees either more the function of management or more that of the managed.

In the capitalistic type of relation, the man who managed an enterprise was also its owner; the owner–manager was the employer. In the new type, on the other hand, a manager is not the owner of a business, nor does he acquire rule over it through investing a large amount of capital. He is, instead, an employee who is held responsible for the efficient and effective operation of the business as a whole, and who is given a corresponding authority to control it. A manager of this type may be called a "professional manager."

There are, of course, even in this stage, many who can claim at least partial ownership of a business in their capacities as investors or stockholders. As the process of dispersion and popularization of stockholding goes on, however, all except a small minority are virtually incapable of sharing, not only in the actual operation of an enterprise, but also in the appointment and dismissal of those whom they entrust with its management. The only aspect of ownership the great majority can still enjoy is that they are entitled to receive dividends. In other words, they are no longer capitalists who have control of their own business.

One may wish to say that the goal of industrial democracy will be achieved only when each rank-and-file worker has a voice in the appointment and dismissal of management. If this be the goal, however, it will still take a considerable time to reach it. Even so, it is already evident that the nature and competence of employers in the fourth stage are of a new and different quality. The right of managers to control a business is no longer based on their private ownership. The legitimacy of their right to employ, discharge, reward, and punish workers has also undergone a considerable change. Formerly, a man's legitimacy to control a business derived from his prosperity and full ownership. Today, by contrast, a man is entrusted by the public with the responsibility for the efficient and effective operation of a business, and it is this trust which is the basis of his authority.

To be sure, managers even in this stage will endeavor to make profits. However, they do so not alone for the sake of the investors; it is also required of them for the sake of the employees as well as of consumers in general. Management today assumes a variety of social responsibilities—toward the profits of investors, toward the interests of business connections, toward the benefits of consumers in general, toward the development of local communities, toward the economic growth of the nation, and particularly toward the welfare and self-realization of employees.

Corresponding to this transformation on the part of management, the nature and social roles of general employees have also changed considerably. True, they are still on the side of the managed. Workers in this fourth stage, however, are not a mere labor force, nor are they necessarily a proletariat. Although the disparity between rich and poor, as well as class distinctions, still remain in most advanced countries, the class hatred the working masses used to have toward the power elite is now being replaced by their aspiration for, and self-identification with, the middle classes.[9]

No doubt, in the majority of industrial establishments, the workers are not yet fully treated as partners to management. However, as is evident in the New Year's addresses in the beginning of this chapter, management today is coming to be more in need of the workers' help. Without their willing cooperation, it will be practically impossible for a manager to accomplish his mission. Moreover, managers today are, in most cases, themselves employees. And this fact indicates that people who are now customarily called "employees" are in reality no longer a labor force, nor followers, nor servants, but are managers' partners, colleagues, or fellow workers, and therefore deserve to be treated as such.

This treatment, moreover, should not be confined to those at the higher echelons of the company. Ordinary operators working at the bottom, too, are entitled to treatment as partners to management, since they voluntarily entered the firm and have been working there faithfully for several years. They are management's equals in that both are employees of the organization. In fact, most were originally those who failed to become staff officials, owing to their insufficient educational background. Whether a department head, a first-line supervisor, or a shop-level operator, all the employees in a firm are its management's partners and are held responsible, at least partially, for the successful operation of the firm. Their right to participate in managerial decisions is nothing but the other side of this responsibility.

9. There are a number of research data to show the recent change in class identification of the Japanese working masses. According to our studies in the attitudinal changes among the Tokyo Metropolitan males conducted in 1955 and 1960, for example, the proportions of those who identified themselves with the "middle classes" were 23 per cent in 1955 and 29 per cent in 1960, while those identifying themselves with the "working classes" were 74 and 62 per cent, respectively. See Kunio Odaka, "The Middle Classes in Japan," *Contemporary Japan,* 28, no. 1 (September 1964), pp. 22–25.

So far I have outlined the development of democracy in employer–employee relations. At least in its last two stages, the development is inseparable from the democratization of the other aspect of industrial relations, that is, union–management relationships. In brief, this is the process in which the labor union is becoming a body equal with management by gaining more voice in negotiations and conferences with management.

The germ of labor unions can be found already in the second stage of the development of employer–employee relations. It is, however, in the *third* stage that they appear on a full scale. As was seen above, workers in this stage were "alienated wage slaves" in their work place and "a lifelong oppressed class" in their social life. Employers, on the other hand, were capitalists who monopolized the right to control a business, and were quite indifferent to the workers' welfare. As a consequence, workers had to unite against capitalists in order to improve their condition. The labor union originated from such enforced necessity on the part of workers.

The labor union, therefore, was in the first place an opposition group, whose purpose was to protect the interests of workers who were weak in their relations with employers. It was secondly a struggle organization, whose purpose was to attempt class strife in order to alter the destiny of workers as lifelong subordinates. For the workers it was possible only through labor unions to better their lives and status. It is understandable that the workers identified themselves deeply with a union, and that they considered the organization as their base for most social activities.

Such functions of the labor union remain unchanged in the *fourth,* or present, stage of development. It is its primary mission even today to be in opposition to management, and to raise the workers' status and welfare through collective bargaining and struggle. In this sense, it might be a cunning plot for management to propose that unions be called its "partners."[10] Though seemingly friendly, this proposal actually is a denial of the *raison d'être* of labor unions.

This holds true for a Japanese labor union, where most members are employees actually working in a given enterprise. One cannot say that a union is a "partner" of management on the ground that it is mainly composed of employees who should individually be treated as its partners. The functions of a union as an organization and the role of an employee are two different things. Even if a worker is at one and the same time an employee and a union member, these two roles should be distinguished from one another.

Although the primary function of the union as an opponent is unchanged in the fourth stage, there is a growing tendency in most advanced industrial nations for the character of this function to become constructive rather than destructive, critical rather than combative. This is a consequence of the marked increase in the union's voice in such advanced nations. Today unions are no longer looked upon as illegal or anti-social organizations. They are firm social establishments publicly recognized by the state. At the same time, however, it is undeniable that unions today are gradually losing the youthful attraction they used to have.

This transformation of the labor union is symbolized by the fact that the system of joint consultation with management is now pervasive in the advanced countries, including Japan. Collective bargaining and the strike are no longer its only major activities. Nor is the view popular that the union is a strategic base for class struggle or social revolution. Moreover, because of recent remarkable economic growth, the

10. In recent Japan, management is often tempted to call unions its "partners," in the hope that this will calm down union aggressiveness. A spokesman of the Keizai Doyukai, the leading Japanese management organization, for instance, recently advocated that, since Japanese labor unions are mostly enterprise unions, they should be treated as management's partners.

labor union can no longer claim to be the only route by which workers can improve their lives. Workers in recent years are coming to realize that the betterment of their livelihood is directly affected by the success of the company to which they belong, as well as by the industrial prosperity of the nation as a whole. As a result, the union as seen by workers today is becoming more and more a sort of service agency from which they can derive benefits in exchange for union dues.[11] Most members nowadays tend to feel themselves "beneficiaries" of a union, rather than "fighters" for labor.

Thus, the fourth stage in the development of industrial democracy, in which we now find ourselves, is characterized first by the appearance in large numbers of professional managers and the corresponding transformation of workers into their partners. Second, it is characterized by the increasing power of labor unions and the concurrent change in their social functions.

It is to be admitted that these changes in industrial relations are not yet the general rule even in the advanced countries. The remnants of the capitalistic type of industrial relations we have seen in the third stage are still strong in many countries, perhaps still stronger than any other types. It is also to be admitted that there still remain in many advanced countries such undemocratic facts of life as the increase in the alienation of white-collar workers, the existence of the unemployed and the underemployed, and the domination of large-scale over small-scale industries. Nevertheless, it is undeniable that the new type of industrial relations is now increasing in power year by year, and this fact must not be lost to sight. In Japan at least, this has become evident since the end of World War II, and particularly since the second half of the 1950s.

11. According to Arthur M. Ross, such a transformation is now rather common in the United States. See Arthur M. Ross, "The New Industrial Relations," in *Changing Patterns of Industrial Relations,* comp. Japan Institute of Labour (Tokyo: Japan Institute of Labour, 1965), pp. 156–160.

Institutionalizing Workers' Participation

In the previous section, I have tried to show that democratizing industrial management through a system of workers' participation is a task of historical necessity for management today. What steps, then, should management take to institutionalize workers' participation? What procedures should workers follow in order to carry out the system?

It is to be emphasized in this connection that the system of worker participation should be institutionalized in the formal organization of the firm as a whole. An enterprise, of course, consists of a number of departments and sections which, in turn, comprise at least three different levels—the top, the middle, and the shop. The method of practicing participation can be different in each department or at each level. It is of importance, however, to establish an overall system of spontaneous participation by employees, by coordinating each of the different methods used under a guiding principle of industrial democracy. In the following pages, I shall first describe various methods of worker participation, and then outline my view of the steps to be taken for its institutionalization, steps which I believe will be successful.

Methods of Workers' Participation. As was discussed earlier, the opportunities for participation should be first given to those at the bottom, or the shop, level of a firm. The nature and scope of the decisions in which the shop-level workers are to participate will necessarily be confined to matters within their own workshops. Though to a different degree, the same is true for the middle-level workers. The decisions in which they can take part will for both be confined to the scope of what has been decided, in the main, by top management. On the other hand, more concrete decisions relating to detailed plans and regulations within each work place should be left to those at the middle and shop levels.

The items to be decided and carried out at these levels may be classified into the following categories:

1. Jobs and work regulations—for example, decisions on procedures for daily work in the shop; drafting of a production schedule; planning for increased efficiency; decisions about the regulations with regard to work speed, rest pauses, shift and rotation of work, and overtime work; and self-government in the execution of these decisions and plans by employees.

2. Discipline and human relations—for example, disciplinary problems within the work group; planning for improved communications; devising the means to heighten employees' work morale; decisions to democratize supervisory practices; and self-government in the execution of these reforms and plans by employees.

3. The facilities and institutions of the company as a whole —for instance, improvement of the company's safety and sanitary facilities; reform of its welfare facilities; planning for a new system of physical and cultural education for employees; and self-government of these matters by employees.

In addition, we may wish to add the following two categories: working conditions, including wages and salaries, the wage system, allowances, fringe benefits, as well as working hours, and paid holidays; and personnel affairs, such as employment, placement, promotion, punishment, layoff, and discharge. It would be advisable, however, to exclude these two categories from the scope of the decisions in which employees can take part, at least during the early development of the system of employee participation. The category of working conditions includes those items that are primarily to be treated by labor unions through collective bargaining, while if employees were to share in decisions about personnel affairs, they would be likely to interfere in one another's individual interests.

This should not, however, prevent an employee from expressing his desires as to his own transfer and promotion. He should also be encouraged to make suggestions about such company personnel practices as the systems of examination and personnel evaluation.

I want to add a word in this connection as regards the opinion that the selection of first-line supervisors, such as foremen, should be made by rank-and-file operators who are to work under their supervision. G. D. H. Cole, the noted advocate of guild socialism, for example, asserts that it is not sufficient for workers to be treated as management's partners; they should also be given the right to elect their own supervisors. According to him, this is essential for industrial democracy.[12]

No doubt, this practice is democratic. Particularly those who look for a model of industrial democracy in the representative system in politics will highly applaud it. In my opinion, however, it is not necessary for supervisory personnel to be elected from among the workers for the organization to be called "democratic." As often happens in politics, if the supervisors elected in this way do not want to let the workers take part in decisions about the operation of the workshop, there will be no democracy. Moreover, under present circumstances, supervisory people themselves, not to mention top management, will surely object to such a practice, so long as it is adopted as a formal rule of the organization. In fact, Cole himself thinks that the institutionalization of this practice will be possible only when it is supported by the enactment of state legislation.

So much for the areas of managerial function in which rank-and-file workers may participate. What, then, are the

12. G. D. H. Cole, *The Case for Industrial Partnership* (London: Macmillan, 1957), chaps. 4 and 7.

methods or procedures by which they are to take part? Generally speaking, these may be grouped into the following two categories:

Indirect participation, in which a certain number of employee representatives elected from among all the workers of an industrial establishment take part in managerial decisions. The employees do not, however, directly participate in the planning, goal setting, and other decision-making processes stated above.

Direct or *full* participation, in which all the employees of an establishment are to take part not only in the decision making, but also in the execution of the decisions about the operation and management of their own workshops.

The first category has the advantage that employees are able to share, even though indirectly, in the highest level of managerial decisions on the operation of the establishment as a whole, since their representatives usually are invited to hold joint consultations with top management. Rank-and-file workers, however, cannot directly share in any decisions, not only at the highest level but even as regards their own workshops. As a result, they may feel themselves deprived of autonomy in, and responsibility for, their own work. This will more likely happen as the establishment grows in size, and the social distance between the representatives and the rank-and-file increases.

The disadvantage that may attach to the second category is that the decisions in which employees can in general share are limited to those at the lower levels. They usually have little chance to participate in the planning and goal setting of the whole enterprise. By means of this type of participation, however, rank-and-file workers can have a real chance to govern themselves in making and executing decisions as to the functioning of their own work places.

If all the workers employed in an organization are to be treated as partners of management, and if they are entitled to be their own bosses at work, their participation should be of the direct type. It would be desirable, however, that some form of indirect participation be used together with it, so that the disadvantages that may accompany direct participation can be overcome.

There are two alternatives when the second category is adopted; that is, direct participation may be either *individual* or *collective*. In the former, employees share individually and separately in managerial decisions. The suggestion system, for example, is a participative practice by which individual workers contribute to decisions by making suggestions for improvement of work efficiency, planning for an organizational change, and so on.

In collective or group participation, on the other hand, workers as members of a work team collectively make decisions as to the operation of the team itself, which in its turn shares in managerial decisions at higher levels. The systems of "project teams" and "autonomous small groups," to be described later, are good examples.

Turning now to more concrete procedures for employee participation, we may classify them into four groups. Though only inadequately, at least the former two of these procedures are already in operation in countries of free enterprise, including Japan. They are:

Participation by suggestion,

Participation by consultation,

Participation by co-determination,

Participation by self-government.

1. Participation by *suggestion* is a form in which individual employees, in most cases separately, try to influence managerial decisions by making suggestions as to technological improvement, increases in work efficiency, the betterment of human relations, or their own transfer and promotion. The suggestion scheme, the "speak-up" program, and employee opinion surveys are the actual procedures of this type.

The suggestion scheme in particular has a long history, and has been widely practiced in many industrial countries. According to a survey conducted in 1965 by the Nikkeiren (Federation of Employers' Associations), in 1,061 business firms in Japan answering the questionnaire, as many as 72 per cent reported that they had adopted this system.

There are, however, several kinds of suggestion scheme, some of which may not be considered an appropriate method for employee participation. Matters open to suggestion, for example, range from technological questions alone to the improvements of communication, supervisory practices, or organizational structure. The procedures used in judging suggestions also vary; in some cases they are examined by a group of judges representing only management, while in others they are screened by a committee consisting of the representatives of workers as well. The most conventional form is for suggestions containing ideas for improving work efficiency to be screened by a managerial group of judges, with those considered most useful for the company receiving awards.

The latter form, however, has as its chief aim the exploitation of original employee ideas for the benefit of the company. It is not necessarily designed to encourage worker participation. For this purpose, it will be necessary to adopt the practice whereby sincere suggestions, regardless of their usefulness to the company, are fully recognized as contributions to managerial decision making. To offer awards will not be necessary, if all the suggestions judged as contributing to managerial policy are to be reported on a billboard or in the company newspaper. It is desirable that the screening committee be composed of representatives of both management and the employees.

The "speak-up" program is a method of more direct suggestion in which rank-and-file employees can bring directly to top management their opinions about company policy, their desires for their own transfer or promotion, their complaints about working conditions, and so forth. In this program, workers are usually expected to employ a prescribed form. In a similar method, often called the "open-door" policy, on the other hand, workers are encouraged to speak directly to top management about their complaints and desires, management in its turn inviting rank-and-file employees freely to show themselves at their offices.

An employee opinion survey, of course, is not a method by which workers take part in managerial decisions. Nevertheless, in my opinion, it is a necessary initial step for institutionalizing true systems of employee participation. Only on the basis of the data obtained from an opinion survey can management develop a proper design for their institutionalization. At the same time, employees themselves, though anonymously, can express their opinions and criticisms about the states of their workshop and of the company as a whole. As a rule, this kind of survey is conducted by a group of outside experts, using an anonymous, paper-and-pencil questionnaire, applied to a sample representing the employees. The questions generally cover items on job, equipment, working conditions, workshop environment, welfare facilities, supervisory practices, managerial policies, the labor union, and union–management relations.[13]

Based upon the data resulting from an opinion survey, management is able not only to formulate or change administrative practices. It can also allow the employees to take part in the planning of managerial reform by letting them, at each of their workshops, examine and analyze the survey data, and by inviting them collectively to think of counterplans to undesirable tendencies revealed by the survey. This procedure for utilizing survey results, often called "feedback," can be a base for a suggestion scheme, as well as

---

13. The Workers' Allegiance Surveys described in Chapter Four are good examples.

workshop conferences, to be explained below. In the former, employees are encouraged to make collective suggestions, for example, as to an organizational change, while in the latter, they are led to discuss their own problems based upon the survey data. It is desirable that employee opinion surveys be carried out periodically, for instance, at intervals of two or three years.

2. Participation by *consultation* is a method by which workers or their representatives periodically confer with their supervisors or other management representatives about planning and goal setting, as well as about improvements and reforms, with regard to the functioning of a workshop or of the establishment as a whole. Although in such cases workers are not entitled to make final decisions themselves, they are encouraged to influence managerial decision making by freely expressing their opinions, desires, and criticism. Workshop conferences, employee committees, and the system of "employee directors" are examples of this type.

Workshop conferences are a practice fairly common in most industrial countries, and according to the Nikkeiren survey quoted above, 58 per cent of the Japanese firms studied had adopted it by 1965. There is, however, a variety of forms of this system, ranging from a fifteen-minute meeting for briefing and communication held every morning among members of a small work unit to a series of two-hour conferences periodically convened after closing hours at each level of the plant. In some cases, even an informal luncheon meeting occasionally held on the initiative of a section chief is called a "workshop conference."

Essential conditions for the workshop conference as a participative device are that it should be convened at least twice a month by the workshop leaders; that it should be attended by all members, divided if necessary into small groups, each consisting of not more than twenty workers; that it should deal with problems the solution of which is a matter of common concern of the participants; and that it should devote most of its time to discussion by rank-and-file workers. It is not desirable to use the conference as grievance machinery. To my way of thinking, grievances are basically personal affairs and can best be remedied through some personal channels, for example, personal consultation with a supervisor or a suggestion box. The "collective" grievances most workers in a firm have about their working conditions, on the other hand, should be handled through their labor union.

In my opinion, one of the most effective uses of the workshop conference is to let workers discuss their problems on the basis of the data obtained from an employee opinion survey. Since such a survey usually reveals problems of common concern, workers will automatically be interested in discussing problems. It is also possible to utilize reliable information reported in the company newspaper or in the organ of the labor union as topics for discussion.

Employee activities through various committees can be another form of participation of this type. Such committees are organized for the purposes, for example, of setting new goals of work efficiency, of planning for an organizational reform, or of improving employee welfare facilities. In most cases, a committee consists of a limited number of qualified employees selected and appointed by the company, with a few members representing management. It is desirable, however, that members of a committee be elected or nominated, as far as possible, by rank-and-file employees. It is also desirable that a progress report of each committee be prepared and made known to employees at large.

Important though it is as a method of employee participation, the committee system has its inevitable limitations. For one thing, employees joining committees are necessarily limited in number. For another, shop-level semiskilled workers are only rarely selected for a committee, unless it is specially organized for solving shop-level problems. Moreover, most committees last for only a short period, usually breaking up after they have carried out a specific task.

Another example of worker participation of this genre is the system of "employee directors," in which a limited number of representatives elected or nominated by employees at large are invited, as part-time directors, to attend meetings of the board of directors of the company. Though in theory they can share in managerial decisions at the highest level, actually they are, in most cases, treated merely as consultants or advisers. Moreover, there are even today a good many company executives who consider the system an infringement of managerial prerogatives. Labor unions, on the other hand, are apt to regard the system as a threat to the existing machinery for joint consultation between union and management. According to the Nikkeiren survey quoted above, only 4 per cent of the Japanese firms studied in 1965 had yet introduced this system.

In addition, one may cite as a noteworthy example of this category the *ringi* system, which has been practiced as a unique method of decision making in Japanese business organizations since the early Meiji era. This system adopts a form of group decision making, with written proposals (*ringisho*) being circulated from middle to top echelons of management, so that it has an appearance that is democratic and participative.

The *ringi* system is still commonly found among large-scale enterprises in Japan from mere force of habit. Many progressive business executives, however, are well aware that the system is already obsolete and hinders the rational planning and quick action which are prerequisites to modern management. For one thing, the system is obviously inefficient in that it requires too many steps of authorization before a proposal is approved and brought into operation. Moreover, one should bear in mind that the responsibility for the decisions thus made lies neither with the person who drafts the original proposal nor with those who perfunctorily authorize it. One may even characterize the system as a skillful device for obscuring the locus of responsibility in managerial decision making. Despite its democratic appearance, the *ringi* system actually has little in common with a participative method of management.

3. Participation by *co-determination* is a procedure by which workers or their representatives periodically confer with management and jointly make decisions about the operation of a workshop or of the establishment as a whole. Workers or their representatives, in this case, not only can take part in managerial decisions but also are held responsible for their execution.

There are two kinds of procedure involved here, namely, co-determination by representatives for management and the workers, and co-determination by the entire membership of a workshop, including supervisors. The former practice is best represented by the system known as "partnership management." This practice, modeled after the German system of *Partnerschaftsbetrieb,* has been introduced recently into several small-scale enterprises in Japan.[14] Unlike the employee director system above, this case uses employee representatives who are actually empowered to co-determine managerial policies of the firm as a whole.

This system, however, has its limitations. Even if it can be successfully adopted in small-sized firms, with fewer employees than two hundred, there is no guarantee that it will work effectively in a large-scale enterprise as well. In the latter case, as often happens in a representative system, employee representatives tend to lose the support of rank-and-

14. Guido Fischer, *Partnerschaft im Betrieb* (Heidelberg: Quelle & Meyer, 1955). One of the successful cases where this practice has been adopted is found at Jujo Paper Board Manufacturing. See Shin'ichiro Tanaka, "The Relationship between Management Behavior and Social Structure" (Paper presented at the Fifteenth CIOS International Management Congress held in Tokyo, November 1969).

file workers, who are likely to become indifferent to their activities. As a result, most workers will have no opportunity, nor consciousness, of participating in the decisions that actually affect them.

In order to avoid these defects, it will be essential for rank-and-file employees to be given ample opportunity to communicate with their representatives. They should, for instance, be provided sufficient information about the questions at issue and have a chance to discuss them with their representatives, before the latter attend a conference with management. Subsequent to each conference, rank-and-file employees should also be fully informed by their representatives as to what has been discussed and the results of the conference.

The other form, co-determination by all the employees in a workshop, including supervisors, can be successfully practiced if the membership of the workshop does not much exceed twenty, and if the decisions they are jointly to make are concerned primarily with the shop-level operation. On the other hand, it would be almost impossible to let all the workers of an establishment of, say, three thousand take part directly in managerial decisions that affect the operation of the whole company. Only through a representative system will they be able to share in such managerial decisions.

4. Participation by *self-government* is a form in which members of a work group are empowered to govern themselves in making and executing decisions about the operation of their group, which in its turn shares in managerial decisions at higher levels. The general framework of policy for the functioning of the work group is usually predetermined by management at higher levels, at which only leaders of the work group can participate. As regards the details of the group's own activities, however, group members themselves, under the guidance of their leaders, are encouraged to make collective decisions and to carry them out on their own re-

sponsibility. Through this type of participation workers can have a real chance to exercise autonomy in their work place. Project teams, the autonomous small group system, and self-government of workers in larger groups are examples of procedure of this type.

Project teams, also called "task forces," are small groups of self-governing workers organized temporarily for the purpose of carrying out a new project or of accomplishing a special task. Members of a project team are selected, on the basis of special abilities required, from among the employees belonging to a variety of sections and departments of the company. Although they are usually selected by higher-echelon management, they can also be elected by fellow employees of each section or department. Unlike the system of employee committees described above, project teams are empowered to make and execute most of the decisions necessary for their task, though the task itself is predetermined by management. Today, a number of Japanese business firms whose management is known to be progressive have introduced this system.

This important practice, however, has its limitations, similar to those pointed out for the employee committee. The fact that project teams are organized out of those specially qualified for a task, in particular, tends to make the practice remote from ordinary workers.

In contrast to project teams, which are temporarily organized for specific tasks, autonomous small groups are basic work units engaged in everyday activities on the shop floor. Each such work unit is a face-to-face group, consisting of fewer than twenty employees. Members of the group are so selected that they make a team of workers with similar qualifications. Each team acts under the guidance of a leader, who concurrently is a member of an upper-level small group. Leaders of such upper-level groups again form another group of a still higher rank, and so on. In this manner, the whole organization of a plant or a firm is made up of a network of vertically and horizontally interconnected small work groups.

The leader of a basic work team is selected from among members of the team itself; in some cases he is even informally nominated by the team members before he is formally appointed by management. In any case, a leader is expected to act primarily as the representative of his group, rather than of management. He is therefore not a boss, but a member of the team who actually works alongside his fellow employees. He is also expected to function not only as the connecting link between lower- and upper-level groups but also as a coordinator between neighboring groups at the same level.

The basic work units are autonomous in that their members are empowered collectively to make and execute decisions about the details of their own activities. Although the broader framework of policy for the operation is predetermined by high-echelon management, each team, through its leader as well as through leaders of upper-level groups, can share in managerial decisions. Each team periodically holds conferences to discuss and decide what the team should do with regard to monthly production goals, work schedules, quality control, control of equipment and fixtures, safety and sanitary facilities, and so forth. Such conferences are occasionally convened jointly by two or more work teams. In addition, each team is entitled to govern itself by making its own rules about its members' absences, tardiness, shifts, overtime, paid holidays, and so on.

Examples of this type of workers' self-government are still scarce not only in Japan but also in other industrial countries. Rensis Likert advocates a similar idea, a new organizational structure based on overlapping groups, which he calls the "participative-group system" of management, but examples of such a system do not seem to be very abundant in the United States.[15] There are, however, a few notable cases in Japan as well as in America, and in these cases the morale and motivation of the workers have become markedly higher as the result of the introduction of this system of employee participation.[16]

In theory, the practice of workers' self-government can be extended to those working in a larger group, like a section or a department of an enterprise. Such a practice is often referred to as the "delegation of power." In actuality, in larger groups the power delegated to rank-and-file employees is, in most cases, only a small portion of the powers given to the chief. So long as there is a disparity in functions between those primarily occupied with actual operations and those who, without operating themselves, are engaged in managerial activities, there will be very little chance of workers' self-government. This functional disparity, moreover, tends to enlarge as the work group increases in size. For in a larger group the social distance between the chief and his rank-and-file workers is greater, and communication and cooperation between them more difficult.

In order to overcome these difficulties, it will be necessary to divide the larger group into a number of small work teams, and to reconstruct it as a network of such teams, following the principles of autonomous small groups described above.

15. Rensis Likert, *New Patterns of Management* (New York: McGraw-Hill, 1961), chap. 14, pp. 222–236; Likert, *The Human Organization: Its Management and Value* (New York: McGraw-Hill, 1967), chaps. 2 and 3, pp. 3–46.

16. One of the examples of the successful introduction of the system of autonomous small groups is found in Japan at a Sony plant, while in the United States, there is a notable case where the failing organization of the Weldon Company was successfully improved through the introduction of what Likert calls "participative-group system" of management. See Shigeru Kobayashi, *Sozo-teki keiei: Sono jissen-teki tankyu* (Creative management: Its practical pursuit; Tokyo: Management Center, 1967), pt. 2, pp. 92–169; Alfred J. Marrow, David G. Bowers, and Stanley E. Seashore, *Management by Participation: Creating a Climate for Personal and Organizational Development* (New York: Harper & Row, 1967), pts. 2 and 3.

Another form of workers' self-government may be practiced in the activities, relating to safety and sanitary facilities or employee welfare and recreational services, of the company as a whole. Although such activities are extracurricular, so to speak, rank-and-file workers can enjoy autonomy in these matters which fairly deeply affect the daily life of the establishment as a whole. Actually, self-government of this type can best be carried out through a special committee organized out of those elected from among workers at large. It is advisable that the membership of the committee be as large as possible so that the majority of workers may have a chance to share in the activities.

A Program for Institutionalization. In the foregoing I have described, together with their merits and demerits, a variety of the more important methods and procedures by which workers can take part in managerial functions. To institutionalize workers' participation, however, it will not be necessary to include all these methods in the overall structure of an organization. What is essential is to pick up as many of these procedures as are suitable for the company, to integrate them into a well-balanced system in its formal organization, and to put it into execution.

Needless to say, each company has its special character and circumstances, and the same set of participative practices will not be selected by each. For example, partnership management through employee representatives will be effective only when the establishment is a small-sized one, with fewer than two hundred employees. If, on the other hand, it is a large-scale establishment, employing several thousand workers, a direct type of participation through, for instance, autonomous small groups will be a basic necessity. Likewise, participation in top-level managerial decisions may prove to be impractical, whether it takes the form of consultation or co-determination, if the decisions usually call for advanced technological training. On the other hand, workshop conferences will always produce a good effect on rank-and-file employees.

Even if a company decides upon a set of suitable practices, however, it will be unwise to try to install them at a single stroke. The program of institutionalization must be put forward gradually, by dividing it into several steps, each emphasizing one or two specific procedures.

As was stated earlier, worker participation should first be introduced into the shop-level activities of an enterprise. To do this, however, it is necessary for top and middle management, as well as shop-level supervisors, to learn the importance of, and the skills in, leadership through delegation of power. Shop-level workers, of course, must undergo a corresponding training for participative practices. The introduction of the practices and the training for them must be put forward side by side. It is not advisable to install an organizational change first and then try to make people adjust themselves to it.

It will be evident, from the foregoing, that the introduction of worker participation presupposes a selection of employees with real ability and aptitude. It also requires that the right men be put in the right places. If, for example, the supervisors and workers who are empowered to govern themselves are incapable or misplaced, it will not only result in a decrease in work efficiency but also will introduce confusion into the workshop. To be sure, under the systems of worker participation, employees will have a real chance to develop their talents and abilities. This is, in fact, one of the major merits of employee participation. However, only through selecting really able individuals and putting them in the proper posts, will the practices be introduced successfully.

The introduction of worker participation is necessarily accompanied by an organizational change, and a change that comes from above or outside without their chance of sharing in it will almost invariably cause resistance on the part of those who are subjected to the change. The institutionalization should not be undertaken in the conventional bureaucratic manner. It should, instead, be carried out by inducing workers to participate in the process of the organizational change itself. The first step thereto will be to inform workers fully of the reasons why the company is adopting these systems. Only through their full understanding of, and willing cooperation in, the institutionalization, will the systems of employee participation become "living" systems, and eventually be internalized in the consciousness and behavior of the workers.

The following is a rough outline program for introducing the systems of employee participation which I have in mind. In this connection, I would suggest that the program should be named a "human development," rather than "worker participation," program, in order to avoid misunderstanding and unnecessary confusion.

The *first* step begins with special training of top- and middle-level managers and staff personnel to give them a better understanding of the fundamentals of leadership through the delegation of power. In addition to a series of ordinary seminars on such topics as the development of industrial democracy, the appraisal of participative in contrast to other types of management, leadership quality and motivation for work, and the like, several sessions of "sensitivity" or "managerial grid" training are given by a couple of outside specialists.[17]

Along with management training, an employee opinion survey is conducted under the guidance of a team of outside

17. For an explanation of "managerial grid" training, see Robert R. Blake and Jane S. Mouton, *The Managerial Grid* (Houston: Gulf Publishing, 1964).

researchers. All the employees below section heads, or, if this proves to be impractical, a large sample representing them, answer a questionnaire designed to elicit their candid opinions about job, equipment, work environment, training and placement, promotion, workshop composition and fellow workers, supervisory practices, participative practices the company has already adopted, and other existing and proposed managerial policies. The questionnaire also includes some of the important problems now confronting the company.

The *second* step starts with a "feedback" of the findings from the opinion survey given to rank-and-file employees. The procedure consists of a series of workshop conferences to discuss the problems at issue, with group suggestions made by workshop members for improving the defects and undesirable usages revealed by the survey.

Shop-level workers, as well as their supervisors, undergo a basic training in participative practices during the process of feedback. The problem-solving approach in workshop conferences is emphasized.

At the same time as the shop-level employee training, the training of middle- and top-level personnel for participative management is continued. The topics of their discussion fully utilize the data obtained from the opinion survey.

On the other hand, and also at this stage, a "human inventory" or an assessment of each of the employees of the firm is made, based upon employees' self-reports, as well as upon the evaluation of their seniors and fellow workers. The assessment must be made primarily in terms of the individual's real ability, inclination, and achievement. The human inventory will be used as fundamental data for a reorganization of the company structure on the basis of autonomous small groups.

In the *third* step, a few work units or sub-sections are set up for an experimental operation of the system of autonomous small groups. The selection of members and the leader of each group is to be made with the utmost care.

At the same time, a company-wide opinion survey is conducted, for the second time but in the same manner as the first, followed again by a company-wide feedback.

The *fourth,* and final, step is characterized by the extension of the principle of autonomous small groups to all the basic work units of the establishment.

Along with this major organizational change, a variety of employee committees and project teams are organized, consisting primarily of the middle-level white-collar workers.

Finally, a group of representatives is elected from among the employees at large and, after a couple of rehearsed meetings, conferences for joint consultation or co-determination by representatives of management and workers are to be periodically convened.

Most essential, in my opinion, for success throughout this series of steps is that top management has a firm belief in, and goodwill toward, worker participation. If top managers have the view of workers that the industrial psychologist Douglas McGregor calls "Theory X"—a view that they are basically lazy, do not like to work unless forced to by threats or monetary incentives, and do not want to take any responsibility for their own work—it is unlikely that any attempt to institutionalize employee participation will be successful.[18] Without doubt, mutual confidence between management and workers is a prerequisite. However, if the attempt begins with even a little mutual confidence, on the basis of which management decides to launch a system of worker participation, I am convinced that this confidence will grow and be strong enough to sustain the operation.

18. Douglas McGregor, *The Human Side of Enterprise* (New York: McGraw-Hill, 1960), chap. 3.

On the Effects of Workers' Participation

The final matters I should like to discuss are some major positive effects to be expected from a system of workers' participation. For convenience, the effects will be treated under the following three heads: those upon workers themselves; those for management or the company; and those for the labor union.

The Effects upon Workers. As for the beneficial effects that rank-and-file employees will gain from the introduction of such a system, the following may be pointed out:

1. Worker participation will increase the workers' right to speak and improve their status in the company. The fact that management has recognized their right to participate already shows an increase in their power. They are no longer to be considered management's "servants," almost a lower grade of human beings. They are instead management's partners, who can act and decide independently.

2. Sharing in decisions will give workers a real opportunity to display and develop their talents and abilities. Employee participation presupposes that men of competence be selected and promoted. On the other hand, it helps actualize employees' potential capabilities by making them share in a variety of managerial functions. In fact, participation will mean human development for the workers.

3. Such a system will also give workers an awareness that they are their own bosses at work. This is particularly true when they are empowered to govern themselves. They no longer feel alienated. Moreover, they will come to take pride in their own work.

4. As a result of items 2 and 3 above, employee participa-

tion will give workers intrinsic satisfaction in, and motivation for, work. They will find their lives at the work place worth living and, as a consequence, they will be able truly to enjoy their leisure.

I know that there are some who do not agree with my view, or who consider it to be too optimistic. I also admit that there are workers for whom the effects I have described will not necessarily prove to be advantageous.

For example, at the same time that it improves human development, workers' participation will also result in keener competition among employees. Those who are able and diligent will be duly selected and promoted, whereas incompetents and idlers will be unsparingly disregarded or downgraded. Those who are likely to fall into the latter category naturally will dislike a system of participation. Such a system also demands that employees make decisions for which they are held responsible. Some employees, however, do not like to accept such responsibility. They would prefer to remain "robots" or "small cogs in a vast machine."

While it is true that there are employees who, for the reasons described, will not like to see practices of worker participation introduced, they are certainly in the minority. The majority will receive some advantage from them and will therefore support their introduction.

At least a couple of research data testify to this point. Recently a nationwide employee opinion survey was conducted in Japan, under the auspices of the Ministry of Labor, at fifty-three plants representing twelve different industries. According to its results, as many as 80.3 per cent of a total of 2,451 workers engaged in automated or semi-automated jobs held favorable views of employee participation. The reasons they gave were that practices of participation will help strengthen their right to speak in the company and will help develop their abilities. By contrast, only 18.1 per cent of the respondents were of a negative opinion, and for the reason that such practices will place a heavier responsibility on themselves and make their jobs harder.[19]

Similar views on employee participation have been ascertained by research conducted in 1962 in Norway, with a sample of 628 blue-collar non-supervisory workers drawn from eighteen establishments in Oslo. When the respondents were asked if they desired to participate more in managerial decisions, over half, or 56 per cent, answered that they would like to take part more in decisions which directly concerned their work, and an additional 16 per cent said that they wanted to participate in more decisions that concerned the management of the establishment as a whole. On the other hand, only 22 per cent answered that they had no special interest in more participation.[20]

Let us in this connection consider a few major arguments which cast doubts upon the introduction of employee participation. The opinions of believers in what McGregor calls "Theory X" are typical. They argue that since workers instinctively avoid making decisions or taking responsibility for them, they will never grow really interested even if they are given ample opportunity to share in the company's decisions. What is worse, workers tend to assume an attitude of suspicion, or even of resistance, whenever management tries to introduce a system of participation, because they suspect it to be a menace to their personal liberty.

Though I cannot be so optimistic as to accept McGregor's other view, namely, "Theory Y," without reserve, in my opinion there is no doubt that "Theory X" is an oversimplified view stressing solely the weak or dark side of human

19. Rodo-sho Tancho Rodo Senmonka Kaigi (Ministry of Labor Committee for the Study of Monotonous Work), "Tancho rodo jittai chosa hokoku" (Report of the [1968] survey of monotonous work; Tokyo, 1969), mimeographed.

20. Harriet Holter, "Attitudes towards Employee Participation in Company Decision-Making Processes: A Study of Non-Supervisory Employees in Some Norwegian Firms," *Human Relations*, 18 (November 1965), pp. 297–322.

nature. True, such weak points will be commonly, and to some extent, found in most men. It is also true, however, that every average person has a certain amount of the brighter elements which are emphasized in McGregor's "Theory Y" —namely, that he has a natural inclination for work, is more willing to work when he is entrusted to do so on his own responsibility, and so on.[21]

It is to be noted, on the other hand, that the workers' reaction is likely to be apathy or resistance if management fails to take proper and effective measures in the course of institutionalizing employee participation. If, for example, management initiates the organizational change so abruptly that workers do not even have time to be well-informed of it, or to understand the reasons why it is necessary for the company, management will almost invariably be met by worker resistance. Again, if management wants to install practices of participation for the purpose primarily of manipulating workers, the latter will never respond to it seriously, or will even try to defeat management's plans by assuming an attitude of apathy or withdrawal.

There is another form of workers' reaction to the introduction of a system of participation. If an organization in which the system is being installed has so far been tightly controlled by an autocratic type of management, workers' initial reaction to the shift from autocratic to participative management is often apathy, resentment, or even open hostility. Likert explains this phenomenon, which can be a shocking experience to management, in terms of "the need to release bottled-up animosity and the need to test the superior's sincerity." He writes: "A company making a major shift in its management practices, therefore, must expect either hostility and aggression or indifference as part of the first stage in the developmental program."[22]

21. McGregor, *Human Side of Enterprise*, chap. 4.
22. Likert, *New Patterns of Management*, p. 245.

In addition, some may object to participation practices on the grounds of certain characteristics of Japanese workers. They will argue that employees in Japan have for so long been accustomed to obeying managerial authority that they will hardly wish to influence managerial decisions by sharing in them. As I pointed out in Chapters One and Three, however, workers' attitudes toward authority have undergone a considerable change since the end of World War II, and the traditional authoritarian type of management has been rapidly losing its hold upon workers. Younger employees in particular are no longer willing to work under this type of control. This is precisely one of the reasons why I consider the introduction of worker participation to be an important task for Japanese management today.

Finally, there is still another, and perhaps more convincing, opinion which is also skeptical about the effects of employee participation. This view assumes that as a result of the general prosperity attained by many industrial countries including Japan since World War II, the major concern of workers in these countries has shifted from their work to their leisure, from their work place to their home. It is argued that today workers are becoming less disturbed than before by frustration and alienation within the sphere of their work. Their jobs for them are coming to be no more than the means by which they earn as much money as possible, so that they can enjoy their leisure more. It is further argued that for this very reason workers can no longer be enthusiastic about the introduction of employee participation.

There is little doubt that workers today, not only in Japan but also in many other industrial countries, have come to be more interested than before in their leisure. As industrialization progresses and society becomes more affluent, workers' earnings steadily increase and their working hours gradually shorten, with the result that they can afford more time and money to spend on leisure. In addition, the number of places of amusement has grown so much that workers have more chance to take their leisure. What is more, they tend to find in their leisure an opportunity to realize themselves. The world we now live in is certainly one of mass leisure.

This does not necessarily mean, however, that workers today have lost their basic interest in work and occupation. Those who believe that their jobs are only a means to earn money for their leisure are actually in a small minority. Though they naturally are not willing to occupy themselves with monotonous routine work, workers do show their sincere interest in a meaningful job adapted to their ability. In Japan at least, this inclination seems even more evident among younger workers than it was in prewar years.

The results of my more recent research conducted at two medium-sized manufacturing firms in Japan, cited in Chapter Five, seem to testify to this point. In these studies, a total of some six hundred and one thousand employees, respectively, of the two firms were asked whether their modes of life were more work-oriented or more leisure-oriented. It was found that while employees who proved to be more work-oriented were in the minority at 19 per cent and 9 per cent respectively,[23] those who answered that they were primarily leisure-oriented were still fewer, at only 5 and 7 per cent. On the other hand, over half, or 51 and 64 per cent, respectively, answered that they were both work- and leisure-oriented at the same time. From this we may conclude that some 70 per cent of the employees were at least partly work-oriented.

That a majority of workers still have a basic inclination for work, however, cannot be a phenomenon peculiar to Japan. Why, then, do workers try to seek more opportunities for personal fulfillment in their leisure?

23. Actually, this category consists of two different types: "work-oriented" and "identity," or a type who thinks of work as one's pleasure. The percentages of these two types were 12 and 7 at one firm, and 4 and 5 at the other. See Chapter Five, section on "Work and Leisure: Five Types of Worker Attitude."

In my opinion, the basic reason lies in the fact that the job and work environment for many workers tend to be frustrating and alienating chiefly because they are overcontrolled by the faceless autocracy of modern organization. This has been so at least since Charlie Chaplin brought his wry humor to the movie *Modern Times* in the 1930s. More recently, along with the increase in size and bureaucratization of industrial organizations, this tendency has intensified. Workers, aware of their inability to reform this bureaucratic over-organization by their own efforts, are compelled to seek self-realization in leisure in order to flee its evil influences.

There is no guarantee, however, that those who are frustrated in their jobs can really satisfy themselves in leisure. Rather, the reverse is the case. Workers who are truly satisfied and happy in their jobs are more likely also to enjoy themselves in leisure. It is erroneous, for this reason, to assume that, as the major concern of workers has shifted from work to leisure, they will be less likely to be disturbed by the frustrating and alienating conditions of the work place.

Only through an organizational reform which introduces a system of worker participation can workers regain personal fulfillment in the work place. This means that once the system is successfully installed, a majority of workers can become enthusiastic for the newly introduced practices. I am sure that employee participation will convert at least a certain proportion of workers who now are primarily interested in their leisure or their home into more work-oriented people.

The Effects for Management. What beneficial effects, then, will a system of worker participation have for management or the company? Although the main constituents of participation are workers, it is management that must give birth to the system as an organizational reform. If management fails to recognize the benefits accruing to itself from employee participation, it will not naturally be willing to initiate the system. Let me, therefore, enumerate these benefits, as follows:

1. As was mentioned, workers' participation will heighten their morale and motivation. As a result, it will create an animated atmosphere within the enterprise. It will also bring about a stabilizing effect upon labor–management relations in the firm as a whole, as well as in each workshop. It is to be remembered, however, that at the initial stage of its introduction, management may experience an adverse reaction from the workers.

2. As a result of heightened morale, participation practices will promote work efficiency and productivity in each workshop, and consequently in the company as a whole. This effect may not follow directly upon their introduction, nor is higher morale necessarily accompanied by higher productivity. Nevertheless, these practices will in the long run make positive contributions to increased work efficiency and therefore to the profit of the company.

3. As a result of the benefits upon workers themselves, participation will help them to develop a new form of self-identification with the workshop and the enterprise as a whole. True, such an identification will be quite different from the traditional pattern of company loyalty; unlike the latter, it will be primarily the attachment of a free individual, who otherwise can move to another company at any time he wishes, to his work and fellow workers. Nonetheless, it will have a "fixing" effect, now keenly sought by management in Japan, upon employees.

4. Finally, participation means human development for workers. It will help bring forward a new type of worker who, in contrast to the "small cog" or spiritless "organization man," will be willing to tackle a tough job which challenges his real ability. This is probably the most im-

portant positive effect management can expect from worker participation. If the main body of employees of a company is composed of men with higher morale, higher efficiency, and higher capacity, this means that the company has gained real value in terms of the creative potential of its human factors.

In spite of these substantial benefits, management has often been hesitant to introduce worker participation. There are several reasons for such indecision.

To begin with, some business executives may reject worker participation on the grounds that it is an illegitimate intrusion upon managerial prerogatives. In America, for example, this type of reaction seems to be still common, though, as I have previously pointed out, it is based on an out-of-date idea.

In the case of Japanese executives, it is more likely that they avoid worker participation because they consider that it is somewhat "socialistic" or that it will encourage labor unions to interfere in everyday operational matters of the company. I admit that worker participation is a "socialistic" idea in the broader sense of the term. It is to be remembered, however, that at the fourth or the present stage of the development of industrial democracy, management in many "capitalistic" countries has necessarily taken up participative practices. In most socialistic countries except Yugoslavia, on the other hand, such practices do not seem to have been very much encouraged so far.

As for management's misgivings about union intrusion, it should be recalled that employee participation is distinct from union participation, and must be put into operation independently from the activities of a union. Some may argue, on the grounds that the Japanese labor union is in most cases an "enterprise union" comprising the majority of the employees of an enterprise, that employee participation will eventually become union participation. This I consider to be an abstract argument based on mere formal logic. As a rule, most major activities of a union, whether it is of the enterprise type or not, are carried out through a limited number of union officials. Union functions in joint consultation with management are not an exception to this rule. Employee participation, on the other hand, is a system basically of direct participation by rank-and-file employees on the shop floor.

Another objection to a system of participation assumes that it will bring such unnecessary disturbances as workshop struggles into an enterprise. This objection, however, is again based on a misunderstanding of the function of the system. Even if a workshop struggle is sometimes initiated by outside agitators, it is basically caused by the dissatisfaction and resentment of employees who are deprived of autonomy in the workshop. In a workshop where members have a real chance to govern themselves, on the other hand, they will volunteer to defend it by rejecting outside agitation and interference.

Still another objection to participative practices finds its grounds in an assumption that they will slow down decisions, which have to be made as soon as possible nowadays. There are, in fact, conservative managers who still stick to the stereotyped notion that autocratic control is most efficient if prompt decisions are necessary. This may be true if management will be satisfied with only a short-term increase in efficiency. If, on the other hand, it wants a long-lasting, high level of morale and motivation among employees, it must rely on a participative type of leadership. Besides, the argument is based on the misinterpretation that employee participation is always a system in which rank-and-file workers share in top-level managerial decisions, which often call for advanced technical knowledge. Without doubt, even if management is to consult only employee representatives before reaching final decisions, the process will take a little more time than if decided by management alone. Through joint consultation, however, management can avoid employee apathy or resistance, which might otherwise occur, and to cope with which would take even more time.

In addition, there is an even more important reason for management's hesitating to introduce participative practices; namely, management's distrust of the employees' capability and willingness to engage in such practices. It is conceivable that employees might make blunders in carrying them out, unless they are sufficiently well trained. This of course does not mean, however, that employees are fundamentally incapable of getting used to a system of participation. Before complaining about employees for their lack of capacity, management should give more active attention to their training for such a system.

As regards the assumption that employees are not sufficiently motivated for participative practices, I have already pointed out how unfounded it is. Even if some employees show indifference or reluctance, this should be taken as due to their lack of understanding and training. Again, as Likert says, those employees who have been under the tight control of an autocratic management tend to be suspicious of the shift to a participative managerial system. To cultivate employee willingness to adapt to the new practices will require a step-by-step approach.

A word may be added as to one traditional characteristic of Japanese employer–employee relations in this connection. Although the introduction of a system of employee participation is a considerable reform for a Japanese business organization, in my opinion it will not be so incongruous to the traditional social atmosphere of business firms as is generally imagined. It has been customary to characterize Japanese industrial management as of the highly centralized authoritarian type. This, I think, represents only one aspect. On the other hand, and at the same time, there has been a tendency for Japanese employers to regard it as important to get employee consensus, at least informally, before final decisions are reached about the more important policies of the company. At the lower levels of an enterprise also, the leader of a workshop, though formally functioning as a one-man boss, is actually encouraged to get the general consent of his followers before he makes final decisions. The American labor economists Frederick H. Harbison and Charles A. Myers once said, characterizing the Japanese form of management, that it is "a unique mixture of highly centralized authoritarianism and democratic-participative management."[24] My appeal for worker participation is to urge the Japanese businessmen to reinforce the latter characteristic of Japanese management in accordance with the requirements of the present stage in the development of industrial democracy.

The Effects for Labor Unions. Finally, I shall discuss briefly the effects of worker participation for labor union activities.

In Japan, the term "worker participation" has been often taken to mean participation by a labor union in management. What I mean by the term is, of course, participation by employees, which I think should be sharply distinguished from the former. To avoid confusion, I have repeatedly emphasized that employee participation has little to do with union activities.

This emphasis, however, may have created the impression that employee participation will result in a certain restriction of union functions. I have no such intention. I am decidedly of the opinion that the union is an important or-

24. Frederick H. Harbison and Charles A. Myers, *Management in the Industrial World: An International Analysis* (New York: McGraw-Hill, 1959), pp. 255–256.

ganization by which workers not only can enlarge their life chances but can also improve their social status.[25]

Nor do I think that a union should remain a passive on-looker to the introduction of employee participation. On the contrary, in my opinion, the union should always take an active part in promoting the practices. For it is an ultimate aim of the union movement to enable workers to enjoy more voice in affairs, to develop their human capacities, and to realize themselves, which they can only do through a successful introduction of participative practices. There are, however, still a good many conservative businessmen who have no intention of democratizing their managerial systems. To induce such businessmen to introduce the practices, it will be necessary for workers to enlist the help of a pressure group like the union. In this sense, unions should take an active role in institutionalizing employee participation.

I know there are a number of people, mostly union leaders or intellectuals who are great supporters of unionism, who have objections to a system of employee participation. To discuss its effects for labor unions, it will be necessary for me first to reply to these objections, which may be grouped into the following three categories: the introduction of employee participation will negate the right of participation by a union, which is one of its important functions, particularly in a country like Japan; it will restrict joint consultation between union and management; and it will reinforce employees' company loyalty, with the result that their interest in the union will be by that much reduced.

Let us begin with the first objection. If the disputant means by "participation by a union" the union practice of joint consultation, I do not think that a system of employee participation will infringe on the union's right of participation in management. If, on the other hand, he means full-

25. See Chapter Four, the last section.

scale participation by the union, namely, a system in which union representatives not only co-determine with management all major company policies, but consequently are held responsible for their execution as well, then I think that this very notion is erroneous. For, in so doing, the union has to depart from its basic social role as the "opponent" of management. The union in this case is going to take sides with management, or has already become a part of management, or else it has transformed itself into a company union. There is also no convincing reason why Japanese unions particularly need such a function. Insofar as participation by employees is successfully carried out, there will be no need for a union to repeat what employee representatives have already done with management.

By saying this, however, I do not mean that unions should have nothing to do with the introduction of employee participation in a company. With powerful backing by a union, the introduction can be carried out with less difficulty.

The second objection is often made by those union leaders who regard joint consultation as an important function of modern unionism. They argue that the promotion of employee participation can be an infringement of union's vested interest, since it will restrict this important function. This argument, however, is based on a misunderstanding on their part, because employee participation and the union's joint consultation are basically compatible. These two practices, though distinct in their functions as well as in their main constituents, should be complementary.

To be sure, the systems of "employee directors" and of "partnership management," the two practices classified as part of employee participation, somewhat resemble the machinery for the union's joint consultation, in that in both cases a limited number of employee representatives confer with management about major company policies. However, employee directors, as the name implies, are those who are to play a role in management. The representatives in the system of partnership management also always attend the conferences in a capacity as the partners of management. By contrast, union representatives who are to sit at a joint consultation table represent an organization whose basic functions are opposing or criticizing company policies.

Moreover, the practices of employee directors and of partnership management are merely a part of the procedures included in a system of employee participation. Workshop conferences and the system of autonomous small groups, for example, are more important elements of the latter, and yet are fully compatible with the practice of the union's joint consultation.

The third objection makes a good point in arguing that employee participation will increase company identification of workers, though such an identification will be quite different from the company loyalty of old-fashioned employees. The assumption, however, that an employee who has a strong company identification will necessarily assume adverse attitudes to the union, though commonly found among union officials, is unfounded. As was shown in Chapter Four, there is more often than not a positive, not a negative, correlation between the company and union allegiance of employees. It is, therefore, groundless to fear that by the introduction of employee participation, workers' support of the union will be undermined.

In my opinion, what is more serious for union leaders, and particularly for those in Japan today, is the fact that there has recently been a gradual increase in the number of workers who not only are dissatisfied with their company but also are apathetic toward their union. As industrialization progresses and the living standard of the working masses is improved in many industrial countries, including Japan, workers' identification with their union tends to deteriorate. It will be a pressing need for union leaders today to install some type of participative practice in the union itself so that rank-and-file members will be more interested in union activities.

Viewed in this way, it will be evident that a system of employee participation cannot be a restriction to union functions. Far from it; rather, it will have several positive effects for a union, as follows:

1. By relegating the function of cooperation with management to employees qua employees, rather than union members, unions can clarify their original role and reinforce their basic function as the opponent of management.

2. As a result of this improvement, unions will be immune from becoming "company unions," a stigma often attached to union locals in Japan.

3. With most intra-enterprise activities aiming at production of the "pie" taken over by a system of employee participation, unions will come to concentrate more upon such matters of distribution as wages, welfare facilities, and the like. They will also be able to enlarge the scope of their activities beyond the boundaries of the enterprise and to devote themselves to industry-wide collective bargaining and joint consultation. The limitations from which Japanese enterprise unions have thus far suffered will in this way be gradually overcome.

4. On the other hand, and at the same time, unions can acquire a new prestige as the promoter and defender of employee participation.

We can thus say that a system of employee participation will have beneficial effects upon both workers and management without damaging labor unions. In any event, the time has come to put the system into operation if we are to meet the needs of industrial democracy and of society as a whole.

In conclusion, I should like to add an even broader outlook. The need for worker participation is not confined to industrial organizations. With the increase in size and the degree of bureaucratization of almost all kinds of modern organization, a system of participation is vital wherever people tend to be frustrated and alienated. The recent outbursts of dissatisfaction among university students are not unrelated.

Nor are the similar reactions of workers in the over-controlled jobs and work environment of government offices, religious associations, hospitals, social work organizations, and labor unions. As participative practices and workers' self-government are brought to fruition in these modern organizations, we shall in the long run see an overall democratization of the socio-economic system of the country as a whole.

Bibliography
Index

Bibliography

Abegglen, James C. *The Japanese Factory: Aspects of Its Social Organization.* Glencoe, Ill.: Free Press, 1958.

Argyris, Chris. *Integrating the Individual and the Organization.* New York: John Wiley, 1964.

Barnard, Chester I. *Organization and Management.* Cambridge, Mass.: Harvard University Press, 1948.

Bendix, Reinhard. *Work and Authority in Industry: Ideologies of Management in the Course of Industrialization.* New York: John Wiley, 1956.

Blauner, Robert. *Alienation and Freedom: The Factory Worker and His Industry.* Chicago: University of Chicago Press, 1964.

Blumberg, Paul. *Industrial Democracy: The Sociology of Participation.* London: Constable, 1968.

Brown, William. "Japanese Management—The Cultural Background." In *Culture and Management: Text and Readings in Comparative Management,* ed. Ross A. Webber, pp. 428–442. Homewood, Ill.: Richard D. Irwin, 1969.

Chalmers, W. Ellison, Margaret K. Chandler, Louis L. McQuitty, Ross Stagner, Donald E. Wray, and Milton Derber. *Labor–Management Relations in Illini City.* 2 vols. Champaign, Ill.: University of Illinois Institute of Labor and Industrial Relations, 1953–54.

Chinoy, Ely. *Automobile Workers and the American Dream.* New York: Doubleday, 1955.

Chusho Kigyo-cho (Smaller Enterprise Agency). *Zu de miru chusho kigyo hakusho* (Graphical white paper on smaller enterprises). Tokyo: Doyukan, 1973.

Clegg, H. A. *A New Approach to Industrial Democracy.* Oxford: Basil Blackwell, 1960.

Bibliography

Coch, Lester, and John R. P. French, Jr. "Overcoming Resistance to Change." *Human Relations,* I (August 1948), pp. 512–532.

Cole, G. D. H. *The Case for Industrial Partnership.* London: Macmillan, 1957.

Cole, Robert E. *Japanese Blue Collar: The Changing Tradition.* Berkeley, Calif.: University of California Press, 1971.

Cox, Robert W., Kenneth F. Walker, and L. Greyfié de Bellcombe. "Workers' Participation in Management." *International Institute for Labour Studies Bulletin,* no. 2 (February 1967), pp. 64–125.

Drucker, Peter F. *The New Society: The Anatomy of the Industrial Order.* New York: Harper & Brothers, 1950.

———— *Asu o keieisuru mono* (Managing tomorrow). Tokyo: Nihon Jimu Noritsu Kyokai, 1960.

Dubin, Robert. *The World of Work: Industrial Society and Human Relations.* Englewood Cliffs, N.J.: Prentice-Hall, 1958.

Dumazedier, Joffre. *Toward a Society of Leisure.* Trans. Stewart E. McClure. New York: Free Press, 1967.

Fischer, Guido. *Partnerschaft im Betrieb.* Heidelberg: Quelle & Meyer, 1955.

French, John R. P., Jr., I. C. Ross, S. Kirby, J. R. Nelson, and P. Smyth. "Employee Participation in a Program of Industrial Change." *Personnel,* 35 (November–December 1958), pp. 16–29.

Friedmann, Georges. *Industrial Society: The Emergence of the Human Problems of Automation.* Trans. Harold Sheppard. Glencoe, Ill.: Free Press, 1956.

Golden, Clinton S., and Virginia D. Parker, eds. *Causes of Industrial Peace under Collective Bargaining.* New York: Harper & Row, 1955.

Goldthorpe, John H., David Lockwood, Frank Bechhofer, and Jennifer Platt. *The Affluent Worker: Industrial Attitudes and Behaviour.* London: Cambridge University Press, 1968.

———— *The Affluent Worker: Political Attitudes and Behaviour.* London: Cambridge University Press, 1968.

———— *The Affluent Worker in the Class Structure.* London: Cambridge University Press, 1969.

Haire, Mason, Edwin E. Ghiselli, and Lyman W. Porter. *Managerial Thinking: An International Study.* New York: John Wiley, 1966.

Harbison, Frederick H., and Charles A. Myers. *Management in the Industrial World: An International Analysis.* New York: McGraw-Hill, 1959.

Hazama, Hiroshi. *Nihon romu kanrishi kenkyu* (Studies in the history of Japanese labor and management relations). Tokyo: Diamond-sha, 1964.

Herzberg, Frederick. *Work and the Nature of Man.* London: Staples Press, 1968.

Holter, Harriet. "Attitudes towards Employee Participation in Company Decision-Making Processes: A Study of Non-Supervisory Employees in Some Norwegian Firms." *Human Relations,* 18 (November 1965), pp. 297–322.

Inkeles, Alex. "Industrial Man: The Relation of Status to Experience, Perception, and Value." *American Journal of Sociology,* 66 (July 1960), pp. 1–31.

International Institute for Labour Studies. "Workers' Participation in Management: A Review of Indian Experience (No. 1)." *International Institute for Labour Studies Bulletin,* no. 5 (November 1968), pp. 153–187.

—— "Workers' Participation in Management in Poland (No. 2)." *International Institute for Labour Studies Bulletin,* no. 5 (November 1968), pp. 188–220.

—— "Workers' Participation in Management in France: The Basic Problems (No. 3)." *International Institute for Labour Studies Bulletin,* no. 6 (June 1969), pp. 54–93.

—— "Workers' Participation in Management in the Federal Republic of Germany (No. 4)." *International Institute for Labour Studies Bulletin,* no. 6 (June 1969), pp. 94–148.

—— "Workers' Participation in Management: A Review of United States Experience (No. 5)." *International Institute for Labour Studies Bulletin,* no. 6 (June 1969), pp. 149–209.

—— "Workers' Participation in Management in Israel (No. 6)." *International Institute for Labour Studies Bulletin,* no. 7 (June 1970), pp. 153–199.

—— "Workers' Participation in Management in Japan (No. 7)." *International Institute for Labour Studies Bulletin,* no. 7 (June 1970), pp. 200–251.

—— "Workers' Participation in the Management of Undertakings in Spain (No. 8)." *International Institute for Labour Studies Bulletin,* no. 7 (June 1970), pp. 252–285.

—— "Workers' Participation in Management in Yugoslavia (No. 9)." *International Institute for Labour Studies Bulletin,* no. 9 (1972), pp. 129–172.

—— "Workers' Participation in Management in Great Britain (No. 10)." *International Institute for Labour Studies Bulletin,* no. 9 (1972), pp. 173–207.

International Labour Office. *Workers' Management in Yugoslavia.* Geneva: International Labour Office, 1962.

Japan Institute of Labour. *The Changing Patterns of Industrial Relations.* Tokyo: Japan Institute of Labour, 1965.

Jaques, Elliott. *The Changing Culture of a Factory: A Study of Authority and Participation in an Industrial Setting.* London: Tavistock Publications, 1951.

Keizai Kikaku-cho (Economic Planning Agency). *Keizai hakusho: Nihon keizai no seicho to kindaika* (Economic white paper: The growth and modernization of the Japanese national economy). Tokyo: Shiseido, 1956.

—— *Kokumin seikatsu hakusho* (White paper on national living conditions). Tokyo, Keizai Kikaku-cho, published annually.

Kerr, Clark, John T. Dunlop, Frederick H. Harbison, and Charles A. Myers. *Industrialism and Industrial Man: The Problems of Labor and Management in Economic Growth.* Cambridge, Mass.: Harvard University Press, 1960.

Kobayashi, Shigeru. *Sozo-teki keiei: Sono jissen-teki tankyu* (Creative management: Its practical pursuit). Tokyo: Management Center, 1967.

Kolaja, Jiri. *A Polish Factory: A Case Study of Workers' Participation in Decision Making.* Lexington: University of Kentucky Press, 1960.

—— *Workers' Councils: Yugoslav Experience.* London: Tavistock Publications, 1965.

Larrabee, Eric, and Rolf Meyersohn, eds. *Mass Leisure.* Glencoe, Ill.: Free Press, 1968.

Levine, Solomon B. *Industrial Relations in Postwar Japan.* Urbana: University of Illinois Press, 1958.

Likert, Rensis. *New Patterns of Management.* New York: McGraw-Hill, 1961.

—— *The Human Organization: Its Management and Value.* New York: McGraw-Hill, 1967.

Maher, John R., ed. *New Perspectives in Job Enrichment.* New York: Van Nostrand Reinhold, 1971.

Mann, Floyd C., and L. Richard Hoffman. *Automation and the Worker: A Study of Social Change in Power Plants.* New York: Henry Holt, 1960.

Marrow, Alfred J., David G. Bowers, and Stanley E. Seashore. *Management by Participation: Creating a Climate for Personal and Organizational Development.* New York: Harper & Row, 1967.

Marrow, Alfred J., ed. *The Failure of Success.* New York: Amacom, 1972.

Mayo, Elton. *The Human Problems of an Industrial Civilization.* New York: Macmillan, 1933.

—— *The Social Problems of an Industrial Civilization.* Boston: Harvard University Graduate School of Business Administration, 1945.

McGregor, Douglas. *The Human Side of Enterprise.* New York: McGraw-Hill, 1960.

Morse, Nancy C., and Everett W. Reimer. "The Experimental Change of a Major Organizational Variable." *Journal of Abnormal and Social Psychology,* 52 (January 1956), pp. 120–129.

Myrdal, Gunnar. *Challenge to Affluence.* New York: Pantheon Books, 1963.

Nihon Hoso Kyokai Bunka Kenkyusho (Japan Broadcasting Corporation

Cultural Research Institute). *Nihonjin no seikatsu jikan* (The time-budget of the Japanese). Tokyo: Nihon Hoso Shuppan Kyokai, 1963.

Nihon Seisansei Honbu (Japan Productivity Center). *Nihon no chusho kigyo* (Minor enterprises in Japan). Tokyo: Nihon Seisansei Honbu, 1958.

—— *Roshi kankei hakusho* (White paper on industrial relations). Tokyo: Nihon Seisansei Honbu, published annually.

Odaka, Kunio. "An Iron Workers' Community in Japan: A Study in the Sociology of Industrial Groups." *American Sociological Review,* 15 (April 1950), pp. 186–195.

—— "Japanische Arbeiter zwischen Gewerkschaft und Werksleitung." *Soziale Welt,* 5 (1954), pp. 37–51.

—— "Anwendung der Sozialforschung in der japanischen Betriebsführung." *Soziale Welt,* 11 (1960), pp. 83–95.

—— "Sangyo no kindaika to keiei no minshuka" (Modernization of industry and democratization of management). *Chuo Koron,* no. 884 (July 1961), pp. 26–44.

—— *Sangyo shakaigaku* (Industrial sociology). 2nd rev. ed. Tokyo: Diamond-sha, 1963.

—— "The Middle Classes in Japan." *Contemporary Japan,* 28 (September 1964–June 1965), pp. 10–32, 268–296.

—— *Nihon no keiei* (Japanese industrial management). Tokyo: Chuo Koron-sha, 1965.

—— *Shokugyo no rinri* (Work ethics). Tokyo: Chuo Koron-sha, 1970.

—— "Employee Participation in Japanese Industries." Paper presented at the Seventh World Congress of Sociology, International Sociological Association, Varna, Bulgaria, September 1970. Mimeographed.

Odaka, Kunio, ed. *Shokugyo to kaiso* (Occupations and social stratification). Tokyo: Mainichi Shinbun-sha, 1958.

Packard, Vance. *The Status Seekers: An Exploration of Class Behavior in America.* New York: David McKay, 1959.

Paul, W. J., and K. B. Robertson. *Job Enrichment and Employee Motivation.* London: Gower Press, 1970.

Purcell, Theodore V. *The Worker Speaks His Mind on Company and Union.* Cambridge, Mass.: Harvard University Press, 1952.

—— *Blue Collar Man: Patterns of Dual Allegiance in Industry.* Cambridge, Mass.: Harvard University Press, 1960.

Riesman, David, Nathan Glazer, and Reuel Denney. *The Lonely Crowd: A Study of the Changing American Character.* Abridged ed. New York: Doubleday, 1955.

Rodo-sho (Ministry of Labor). *Rodo ido chosa kekka hokoku* (Report of the Labor Turnover Survey). Tokyo: Rodo-sho, 1960.

—— *Rodo hakusho* (Labor white paper). Tokyo: Rodo-sho, published annually.

Bibliography

———— *Tancho rodo* (Monotonous work). Tokyo: Rodo Gyosei Kenkyusho, 1970.

Rodo-sho Tancho Rodo Senmonka Kaigi (Ministry of Labor Committee for the Study of Monotonous Work). "Tancho rodo jittai chosa hokoku" (Report of the surveys of monotonous work). Tokyo: Rodo-sho, 1968 and 1969. Mimeographed.

Roethlisberger, Fritz J., and William J. Dickson. *Management and the Worker.* Cambridge, Mass.: Harvard University Press, 1939.

Roethlisberger, Fritz J. *Management and Morale.* Cambridge, Mass.: Harvard University Press, 1941.

Rose, Arnold M. *Union Solidarity: The Internal Cohesion of a Labor Union.* Minneapolis: University of Minnesota Press, 1952.

Rosenberg, Bernard, and David Manning White, eds. *Mass Culture: The Popular Arts in America.* Glencoe, Ill.: Free Press, 1957.

Sayles, Leonard, and George Strauss. *The Local Union: Its Place in the Industrial Plant.* New York: Harper & Brothers, 1953.

Seashore, Stanley. *Group Cohesiveness in the Industrial Work Group.* Ann Arbor, Mich.: Publications Distribution Service, University of Michigan, 1955.

Sheppard, Harold L., and Neal Q. Herrick. *Where Have All the Robots Gone?: Worker Dissatisfaction in the '70s.* New York: Free Press, 1972.

Shirai, Taishiro. *Kigyobetsu kumiai* (Enterprise labor union). Tokyo: Chuo Koron-sha, 1968.

Sombart, Werner. "Arbeiter." In *Handwörterbuch der Soziologie,* ed. Alfred Vierkandt, pp. 7–12. Stuttgart: Ferdinand Enke, 1931.

Stagner, Ross. *The Psychology of Industrial Conflict.* New York: John Wiley, 1956.

Stagner, Ross, Theodore V. Purcell, Willard A. Kerr, Hjalmer Rosen, and Walter Gruen. "Dual Allegiance to Union and Management: A Symposium." *Personnel Psychology,* 7 (March 1954), pp. 41–80.

Tannenbaum, Arnold S., and Robert L. Kahn. *Participation in Union Locals.* Evanston, Ill.: Row, Peterson, 1958.

Touraine, Alain. *L'évolution du travail ouvrier aux usines Renault.* Paris: Centre National de la Recherche Scientifique, 1955.

———— *La conscience ouvrière.* Paris: Éditions du Seuil, 1966.

Tsuda, Masumi. *Nenko-teki roshi kankei-ron* (The problem of industrial relations based on seniority system). Tokyo: Minerva Shobo, 1968.

Tsusho Sangyo-sho (Ministry of International Trade and Industry). *Waga kuni sangyo no automation no genjo to shorai* (The present state and future prospects of automation in Japanese industry). Tokyo: Tsusho Sangyo-sho, 1962.

Veblen, Thorstein, B. *The Theory of the Leisure Class: An Economic Study in the Evolution of Institutions.* New York: B. W. Huebsch, 1899.

Bibliography

Vitels, Morris S. *Motivation and Morale in Industry*. New York: W. W. Norton, 1953.

Walker, Charles R., and Robert H. Guest. *The Man on the Assembly Line*. Cambridge, Mass.: Harvard University Press, 1952.

Warner, W. Lloyd, and James C. Abegglen. *Occupational Mobility in American Business and Industry*. Minneapolis: University of Minnesota Press, 1955.

Whitehill, Arthur M., Jr., and Shin'ichi Takezawa. *The Other Worker: A Comparative Study in Industrial Relations in the United States and Japan*. Honolulu: East-West Center Press, 1968.

Whyte, William F. *Men at Work*. Homewood, Ill.: Richard D. Irwin, 1961.

Whyte, William F., Melville Dalton, Donald Roy, Leonard Sayles, Orvis Collins, Frank Miller, George Strauss, Friedrich Fuerstenberg, and Alex Bavelas. *Money and Motivation: An Analysis of Incentives in Industry*. New York: Harper & Row, 1955.

Whyte, William H., Jr. *The Organization Man*. New York: Doubleday, 1956.

Yoshino, M. Y. *Japan's Managerial System: Tradition and Innovation*. Cambridge, Mass.: Massachusetts Institute of Technology, 1968.

Zenkoku Kyoiku Kenkyusho Renmei (National Federation of Institutes of Educational Research). *Kinro seinen no seikatsu* (The life of young workers). Tokyo: Toyokan, 1959.

Index

Abegglen, James C., 3, 6–9, 56–57
Age groups, 110–111, 132, 144, 151.
See also Young workers, Japanese
Alienation, 34, 41–42, 45–47, 49, 51,
73, 132, 168, 195, 201
Allegiance types, 90–94; and age, 110–
112; distribution of, 95, 98–102; and
educational career, 112; and position
in company, 112–113; procedure for
identifying, 94, 94n; structural characteristics of, 103, 106–109; and
technological advancement, 113–115
Assembly-line work, 25–26, 34, 38–43,
46–47
Automation, 22, 31–33, 53–54, 73,
108, 129; bright and dark sides of,
33–36; business, 28, 37, 39; mechanical, 28, 37–38; process, 28, 37,
39, 42, 47; workers' reactions to,
37–38, 59, 63
Autonomous small groups, 179, 187–
189, 189n, 190, 193–194, 209
Autonomy, in workshop, *see* Self-
government, workers'

Blauner, Robert, 41–42, 47
Bureaucratization, 24–27, 47, 51, 73,
211
Business-machine work, 39–41, 43

Chaplin, Charlie, 47, 201

Co-determination, 185–186, 194; German system of, 159
Cole, G. D. H., 177
Company loyalty, 11, 15, 115, 128,
202, 207. *See also* Workers' allegiance, to company
Con-Con type, *see* Discontented
Con-Pro type, *see* Unilateral union allegiance

Dehumanization, 20, 27, 35, 47
Democracy, 14, 120, 163–164
Democratization, of management, 18,
20, 158, 166
Discontented, 76–77, 92, 98–102, 108–
109, 111–115, 118–119, 122
Dissatisfaction, workers', 59, 63, 72–73,
152
Drucker, Peter F., 53–55, 78, 162
Dual allegiance, 91, 98–99, 101, 106–
108, 111–114, 116–117, 117n, 119,
126

Educational career, 59–61, 75–77, 79–
80, 112–113, 129, 132, 152
Employee committees, 182–183
Employee directors, 182, 184–185,
208–209
Employee opinion surveys, 16, 179,
181–183, 192–194

Index

Index

Neutral-Neutral type, *see* Nonpartisan

Ninomiya, Sontoku, 137, 147

Nonpartisan, 91–92, 98–99, 107

Occupational mobility, 56–58

One-man production, 48

Open-door policy, 181

Organization man, 68–70, 73, 202

Other-directed, 69–71

Owen, Robert, 26

Packard, Vance, 61

Participation, 17, 82, 120, 157, 160, 163–164, 195–196, 202; by co-determination, 179, 185–186; by consultation, 179, 182–185; direct or full, 178–179, 204; indirect, 178–179; by self-government, 179, 186–190; by suggestion, 179–182. *See also* Employee participation; Workers' participation

Partnership management, 185, 185n, 190, 208–209

Paternalism, 1, 10, 14–16, 18, 20, 81, 167

Pluralistic industrialism, theory of, 3–9

Political party support, 103, 107–108

Positive correlation types, 115–116, 119

Pro-Con type, *see* Unilateral company allegiance

Process-control work, 39–43, 46–47

Project teams, 179, 187, 194

Pro-Pro type, *see* Dual allegiance

Purcell, Theodore, V., 116–117, 117n

Rationalization, 19–20, 24–28, 35, 48

Riesman, David, 69

Ringi system, 184–185

Rose, Arnold M., 119

Round-table system, 47

Russell, Bertrand, 147–148

Self-fulfillment, 26, 44–45, 48, 201

Self-government, workers', 50–51, 179, 186–190

Semileisure, 140

Sense of belonging, 89, 102, 129

Separation rate, 8n, 130n

Sombart, Werner, 166–168

Speak-up program, 179–180

Split type, 148–150

Stratification and Mobility Surveys, 55–56, 66, 131, 134–135

Success, desire for, 67, 131; philosophy of, 66, 137

Suggestion system, 16, 179–182, 193

Technological innovation, 11, 22, 27, 30, 32–33, 35–36, 38, 53–54, 58, 76, 114–115

Touraine, Alain, 34–35, 42

Traditional values, 103, 107–108

Traditionalism, 1–3, 10

Training, employee, 63–64, 80, 163; job-centered, 15; management, 64, 80, 192; in participative practices, 191, 193; at schools, 64–66

Unemployment, 29, 29n, 36, 36n

Uniform development, theory of, 2–3, 6, 8

Unilateral company allegiance, 91, 98, 106–107, 112, 121–123

Unilateral union allegiance, 92, 98–99, 107–108, 113, 122–123

Union participation, 17–18, 157–160, 203, 206–208

Unions, 163, 172–174, 184, 206–208, 210–211; as management's opponents, 17, 86, 159, 208; as management's partners, 173

Union solidarity, 115, 119

Union splits, 123

Veblen, Thorstein B., 136

Warner, W. Lloyd, 57

Weber, Max, 168

Whyte, William H., Jr., 68

Work hard, 106, 137, 147–148 150; philosophy of, 103, 108

Work-leisure dichotomy, 145, 147–153

Work motivation, 45, 128, 132, 152, 189, 196, 202

Work-oriented, 147, 149–154, 200

Worker ethos, 117–119, 126

Workers' allegiance, 89–91; American studies of, 94, 116–117; to company, 90–91, 102, 128; to union, 90–91, 102. *See also* Allegiance types

Workers' Allegiance Surveys, 74–76, 93–95, 98–113, 129–132, 134, 145–146, 149–152, 200

HARVARD EAST ASIAN SERIES